T0201098

Solvable

HOW WE HEALED THE EARTH,
AND HOW WE CAN DO IT AGAIN

Susan Solomon

The University of Chicago Press
Chicago and London

The University of Chicago Press, Chicago 60637
The University of Chicago Press, Ltd., London
© 2024 by Susan Solomon
Published 2024
Printed in the United States of America

33 32 31 30 29 28 27 26 25 24   1 2 3 4 5

ISBN-13: 978-0-226-82793-3 (cloth)
ISBN-13: 978-0-226-82794-0 (e-book)
DOI: https://doi.org/10.7208/chicago/9780226827940.001.0001

Library of Congress Cataloging-in-Publication Data

Names: Solomon, Susan (Atmospheric chemist), author.
Title: Solvable : how we healed the earth, and how we can
do it again / Susan Solomon.
Other titles: How we healed the earth, and how we can
do it again
Description: Chicago ; London : The University of
Chicago Press, 2024. |
Includes bibliographical references and index.
Identifiers: LCCN 2023050476 | ISBN 9780226827933 (cloth) |
ISBN 9780226827940 (ebook)
Subjects: LCSH: Environmentalism. | Environmentalism—
History. | BISAC: SCIENCE / Global Warming &
Climate Change | POLITICAL SCIENCE / Public Policy /
Environmental Policy
Classification: LCC GE195 .S6525 2024 |
DDC 363.7/06—dc23/eng/20231107
LC record available at https://lccn.loc.gov/2023050476

♾ This paper meets the requirements of ANSI/NISO Z39.48-1992
(Permanence of Paper).

Solvable

CONTENTS

PREFACE

I have hope for this planet. My path of more than four decades through research, teaching, environmental negotiations, and public communication makes me confident that we can solve current environmental problems. I have worked on some of the science myself, taught and been taught by wonderful students, participated in the international negotiating halls, and engaged with media and the public. My goal in this book is to convey how these experiences inspired hope in me—and in turn to inspire you.

We were once in grave peril of poisoning our children with lead additives to gasoline, but no country now uses leaded gas. We nearly drove species like the eagle to the oblivion of extinction with careless use of persistent pesticides like DDT, but today's eagle population is soaring like the magnificent birds themselves. City dwellers in Los Angeles and many other places no longer wheeze and choke through incessant shrouds of daily suffocating smog. We created an unexpected "hole" in the ozone layer that protects all life on Earth from DNA-damaging ultraviolet rays from the Sun, but we've stopped producing certain industrial chemicals worldwide, and it is starting to heal. As recently as 2016, we enacted a new global agreement to phase out other chemicals that replaced those ozone-depleters, because of concerns about their contribution to climate change. The chapters of this book each cover one of these subjects. I begin with the story of a personal challenge and adventure that both brought me to the Antarctic and inspired my own reflections on the principles in this book, and then move to that 2016 environmental success—revealing how it bears directly on action (or inaction) for other drivers of climate change. I then offer a diversity of other examples to further show the principles at work. Taken together, the chapters show that we have successfully addressed not just issues we could see with our own eyes but also those where the science was hard to uncover. They also show that we have enacted change not just in fairly small activities like specialty chemical production, but also in the massive automobile and farming industries. And they show that agreements possible in the twentieth century are still possible in the twenty-first.

We very seldom put nature back exactly as we found it, but we have successfully overcome a host of looming environmental threats of our own making. Those stories illuminate the possible paths to success. I discuss their useful analogs—and a few contrasts—in the book's last chapter, whose topic is humankind's greatest challenge: climate change. You're not reading a handbook, but a roadmap. I won't explain the details of a workable future energy system nor its associated legal or economic intricacies. Instead I offer an argument—that how we healed the Earth in past challenges reveals paths for future success.

While understanding the science was necessary to begin work on each of these dangers, it was never sufficient. Each of these journeys involves people—not just known luminaries, such as Rachel Carson, but also unknown citizens who made huge and essential differences. We will meet birders motivated by concern for their local wetlands, mothers worried about their children's health, and gang members defending their communities, all of whose actions led a broader public response. Armed with public support, enlightened politicians such as the under-appreciated legislative genius Edmund Muskie found ways to turn their own good intentions into enforceable regulations. By tracing each piece of the process—public awakening, social action, effective policies, and technological change and innovation—these stories show how we overcome environmental threats.

In every story, the mobilization of citizen-activists, dedicated scientists, and effective politicians bore fruit through the alignment of what I call the "three P's." For environmental action to succeed, a threat must first involve *personal* distress regarding our own individual safety and concern for our loved ones. It must also become *perceptible*, either because we can see it with our own eyes, or we can understand that it is well-founded and clearly serious. When enough people are confident that the available solutions are *practical*, they can fully engage and jointly change the course of history. Each of our successes demonstrates the unstoppable power of people who are sufficiently committed to do the work long after the inspiring speeches have ended.

Achieving practical solutions often means getting industry to develop or adopt new technologies, but as a rule, corporations don't embrace technological change out of the goodness of their CEOs' hearts. That's why I also argue that we need to catalyze technological change. We'll see that happen when we align public pressure to foster effective regulation and realignments of market forces. Sometimes it's a coercion, as when new laws compelled chemical companies to sell different pesticides. Sometimes it's a demand for a technology we don't even have yet, as when the auto industry was forced to develop an all-new catalytic converter to fight smog. I'll call all of these cases "technology steering" throughout—and I argue that it requires popular will to demand those changes in direction, which again puts the hard work back on citizens like you and me.

Young and old alike feel disheartened today, as social media spew bad news about climate change. The American Psychological Association (APA) describes the term *eco-anxiety* as the chronic fear of environmental disaster that comes from observing the seemingly unstoppable escalation of climate change and the associated concern for one's future and that of future generations. Some young couples are choosing not to have children because of their fear for the planet. But doomsday scenarios and the "It's too late, all is lost" refrain can lead to fatalistic apathy, which keeps us from doing the things we need to do now to stop that climate change cataclysm. Teaching offered me a firsthand view of how young people can be passionate about changing the way we treat our planet. I've also learned how little people of all ages know or remember of past progress, and how environmental victories from the last fifty years offer a pattern for solving the myriad challenges still to come.

I'm a chemist by training, and I have chosen to discuss examples of environmental problems of national or global extent that involve dangerous chemicals or materials. But I believe their lessons—the importance of the personal, perceptible, and the practical—appear in other past problems as well. I am confident that their proven path healing the Earth will help us avoid current environmental problems, and shows that we are on the cusp of success in fighting climate change too.

My message to you: We've done it before, and together we can do it again. It's solvable.

1

Ozone Depletion

SAVING OUR SKINS

The most striking thing was the solitude. As the airplane thundered across an unsettled sky toward Antarctica, I stared out the window at the landscape of shifting ice floes and dark ocean. There were no roads, no settlements, no structures of any kind— not even the occasional lonely ship on the south polar seas. As we approached the continent, the last rays of sunlight faded, replaced by a diffuse blue and purple twilight. That's when I realized we really were on our way to the last place on Earth.

Just the year before, I'd been happily sitting at my desk in my warm and cozy office, where I studied stratospheric chemistry using computer models. The kingpin molecule in that chemistry is ozone, a highly reactive gas produced from oxygen that has unique abilities to absorb high-energy ultraviolet light. Everyone who has ever been sunburned knows that too much sun can be dangerous. Many older people know that too much UV is the primary cause of their skin cancers or cataracts; this was certainly true for me. Earth's fragile ozone shield stands between us and oblivion from the Sun's damaging rays and is what first allowed life to crawl out of the protective ocean and walk on land. A "layer" of ozone formed naturally in the stratosphere as oxygen evolved on Earth, some ten to thirty miles over our heads. But human activities can release a range of chemicals that eat away at it. The most damaging of these are compounds containing chlorine and bromine. Scientists had been expecting some ozone depletion due to human use of chlorofluorocarbon (CFC) chemicals—a small percentage, and a hundred years in the future, a "future" problem. Then in 1985 the British Antarctic Survey shocked the world by publishing a scientific paper declaring that an unexpected "hole" had formed in the ozone layer above their station. They reported a 50% loss of ozone, happening NOW, a wake-up call of outsized proportions. Scientists scrambled madly to figure out if the British measurements were correct, and if so, were humans the cause.[1]

That's what led me to that southbound flight in 1986, as part of a team of researchers on our way to gather the data that would provide the first proof that chlorofluorocarbons were more effective at destroying ozone in the extremely cold Antarctic than anywhere else (including the Arctic, which is almost always warmer). The study of atmospheric chemistry had completely missed some critical extreme cold-temperature chemical reactions and therefore vastly underestimated the seriousness of the environmental problem.

The U.S. National Ozone Expedition (or NOzE as we called it) consisted of sixteen scientists from four different research institutions. At age thirty, I was the youngest and the only woman. Yet I was head project scientist, which made me the group's spokesperson and occasional tough decision-maker. It was a bit of a baptism by fire to be thrust into explaining what we were doing and what we found to the news media and ultimately to policymakers. I found it tough, and I made a lot of beginner's mistakes. But I enjoyed it more than I expected. I also gained a deep appreciation for how policy works, and the critical role of public opinion. CFCs and other ozone-depleting chemicals were later banned everywhere under the most successful global environmental treaty the world has ever known (so far), and humanity has already put the ozone hole on track to slowly heal itself. It would take me decades to understand how and why this miracle of environmental remediation happened.

That first flight to the Antarctic was not truly the beginning of my journey—I had researched the atmosphere for about a decade before I headed south, and my own work followed a long history of ozone research. In the late nineteenth century, scientists began to understand how the ozone layer shields us from an otherwise lethal Sun. If you have ever seen a sunbeam pass through a prism or crystal, you have seen how white light turns into a rainbow of color. These colors correspond to different wavelengths of light, with increasing energy from red to violet. Ultraviolet, a wavelength our eyes cannot see, has so much energy that it can damage life, including us. In 1879, French scientist Marie Alfred Cornu measured different wavelengths of light coming from the Sun and could not detect ultraviolet on the ground. He knew the Sun must be emitting ultraviolet, but that some element in the atmosphere must be blocking it. Still, when he tested them, many of the usual suspects—oxygen, nitrogen, carbon dioxide, water vapor—all failed to block ultraviolet light. A few years later, British scientist Sir Walter Hartley determined that the molecule ozone—three oxygen atoms bonded together—could

3

uniquely absorb some wavelengths of light, including ultraviolet. In 1924, another British scientist G. M. B. Dobson built the first scientific instrument to measure the protective amount of ozone above us. Research on understanding the formation and chemistry of stratospheric ozone continued for decades as an exercise in pure science.[2]

Concern about destroying the ozone layer wasn't brand new when the ozone hole opened. Recognition that human activities might damage the ozone layer began in the 1950s and 60s, when several countries began to contemplate building commercial supersonic transport airplanes, craft capable of whisking people from New York to Tokyo in a matter of a few hours. This involved flying high, in the stratosphere, and faster than the speed of sound. The potential for these planes to deplete the ozone layer—along with questions about their profitability and disruptive sound—became part of a vigorous debate. In the post–World War II boom era, Britain, France, the United States, and the Soviet Union were vying for technological superiority on many fronts, and Cold War tensions added to the desire for both profit and dominance. At first the British and French worked independently, but in 1962 they committed to the joint development of a supersonic aircraft that would be known as the Concorde. The United States chose the Boeing corporation to research whether to build a domestic competitor in 1966.

The timing of the supersonic transport plane issue was fortuitous. An American public already interested in environmental issues like pesticides and smog became intrigued by stratospheric ozone. A Citizen's League Against the Sonic Boom grew in six months from nine to over a thousand members who spanned 39 states in protest against the thundering "bang zones" that would bombard millions living under each supersonic's flight path.[3] On top of that, the aircraft's powerful jet engines would spew pollution into the then-pristine stratosphere. Initial scientific study focused on the water vapor that would be released out the exhaust, and whether it could affect the Earth's climate, or deplete its essential ozone shield.[4]

In 1971 a young postdoctoral meteorologist named Paul Crutzen published a scholarly paper suggesting that a new kind of chemistry could strongly affect stratospheric ozone. He was interested in how ozone concentrations change with increasing altitude in the stratosphere. He was also interested in the role of nitrogen oxides—produced naturally as soil microbes break down nitrogen. The reaction was catalytic, meaning a little bit of nitrogen oxide would destroy a little ozone—but then the nitrogen oxide would regenerate leading to a lot of ozone loss. The paper was a revolution in understanding ozone chemistry, one that earned a share of a Nobel Prize in Chemistry twenty-four years later for Crutzen, along with two other scientists. But it wasn't a warning sign—after all, he was looking at a natural source of nitrogen oxide. Although Crutzen didn't think to mention it in his paper, it quickly dawned on other scientists that the exhaust generated by a fleet of supersonic aircraft flying through the stratosphere would include unnatural nitrogen oxide. If there was enough of the exhaust to compete with natural nitrogen oxide sources, then this could pose a threat to the ozone layer.[5]

The public bombarded the US Department of Transportation with questions, and the department started a task force to study the Concorde's potential environmental impacts. They brought chemists and meteorologists together to think jointly about stratospheric ozone for the first time in a scientific meeting in early 1971. Most meteorologists weren't very concerned about ozone loss, and one presented a paper claiming little effect using a mathematical model. But among the chemists in the audience was Harold Johnston, one of the world's foremost experts on nitrogen oxide chemistry. He spent the night after the meeting reading the paper that had been presented and building his own mathematical model. He cranked the numbers (by hand!), and wrote a sixteen-page analysis overnight. At the next day's meeting session, he announced that one of the chemical reaction rates assumed in the study presented the day before was incorrect—by a shocking factor of 13,000. His own midnight calculations indicated that the nitrogen emissions from a fleet of supersonic planes would pose a serious threat. A few months later, Johnston published a paper in the highly respected journal *Science* arguing that the proposed planes might destroy as much as half of the Earth's essential ozone layer. A high-ranking Boeing engineer phoned Johnston, in Johnston's words, to "curse him out," and the man later went as far as protesting to the National Science Foundation in an apparent attempt to interrupt Johnston's research funds.[6]

To President Nixon, the research meant that his pet project was under attack by a new enemy, and it reportedly made him and his White House staff angry enough to talk internally of "getting this guy Johnston," perhaps by finding a way to cut off his research funding. But Johnston kept getting funded, and remained relentless in his push to get the science out. As he put it, "If it turns out that what you're doing has an impact, you've got to talk about it. You've got to say it."[7] It's just part of the job of doing environmental science that your work might sometime ruffle feathers, even presidential ones.

From the scientific point of view, that meeting was a turning point because it brought a new and powerful scientific community into the ozone depletion story: the chemists. They were influential because chemistry is a big field. It is one of the cornerstones of the scientific enterprise, along with physics and biology. Every major university has a chemistry department but only a few have meteorology. Meteorology is a far younger field, and much smaller, so the arrival of the chemists on the scene not only brought a different and needed perspective, it also brought new ideas, more exposure, and a good deal of scientific clout. The two fields together were far stronger in dealing with the challenges of the public, the policymakers, and the skeptics that were to come than either would have been on its own. I became a chemistry graduate student at Berkeley in 1977, drawn to study atmospheric chemistry under Harold Johnston's guidance. Over the years Johnston and Crutzen had become friends, and I did my PhD thesis working largely with Crutzen, with Johnston's input. I was lucky because of their brilliance and generosity as advisors, and because (like several other young chemists of the day) I gained a life-long interest in and appreciation for how meteorology and chemistry jointly shape our atmosphere.

I was much too young and naive to understand at the time, but looking back years later at the forces that stopped supersonic planes from taking flight in the United States is eye-opening. Many in Congress thought the project too expensive—leading to a tug-of-war familiar to anyone who follows current US politics. Nixon and his successor Gerald Ford repeatedly tried to fund the plan, and the House and Senate invariably struck the money out of the federal budget. Finally, Boeing concluded that they didn't want to build the plane—the slower, conventional 747 with its cheap seats for more people was projected to be a better business bet. The American public's loud opposition to the plane's sonic boom was also getting harder to ignore. Scientific information sometimes offers convenient truths for decision-makers in industry and government. In the midst of all the mess, the news that the ozone layer might be at risk came as a convenient truth that may have been the last nail in the coffin. By mid-1971, the US supersonic transport was a dead duck.[8]

But the biggest threat to the ozone layer wasn't the nitrogen oxides. In the early 1970s, researchers used a clever new technique to analyze air samples taken on a research ship that cruised from Britain to Antarctica. The new method, called electron capture gas chromatography, is exquisitely sensitive, and can detect the presence of gases in amounts a million times smaller than those detected by previous approaches. Throughout the voyage, researchers measured the industrial chemical chlorofluorocarbon-11 (CFC-11, $CFCl_3$). There is no natural source for this chemical, but manufacturers loved to use it as a cheap and efficient propellant in spray cans. Its production had grown exponentially over the previous several decades. Although almost all the spray cans were in the Northern Hemisphere, the researchers detected lots of the chemical in the Southern Hemisphere too. The author of the paper reflected on the values measured all the way from the United Kingdom to the tip of Antarctica. This meant this CFC variety had to be nearly chemically inert—meaning it doesn't react and therefore isn't destroyed as it travels in the lower atmosphere. This is profoundly different from the smog from Los Angeles, which may get as far as neighboring states but never all the way to the tropics, much less Antarctica. That's because smog doesn't live very long in the atmosphere. It can be rained out, and it reacts with vegetation and other surfaces.

The paper noted that the chemical "constitutes no conceivable hazard" and suggested that measuring it might be a great way to learn about how winds move air around the world, a benign "tracer" of air motions.[9] This was not the first time that the apparent inertness of the CFCs was lauded as a wonderful feature. At a meeting of the American Chemical Society in 1930, industrial chemist Thomas Midgley Jr. inhaled some CFC and then exhaled it over a candle to show that it was safe to breathe and so non-flammable that it could even put out a fire, making it a wonderful candidate for use as a safe refrigerant.[10] Midgley's other infamous contribution to environmental chemistry was the industrial development of leaded gasoline. Soon CFC-11 and CFC-12 (CF_2Cl_2) were used in refrigerators and air conditioners, in solvents, in medical applications like asthma inhalers, and to make the bubbles in spongy materials in building insulation and Styrofoam. But towering above all these minor uses was the use of CFCs as the propellants in virtually all spray cans. By 1970, spray can use made up 75% of the US consumption of CFCs.

In 1973, a young Mexican post-doc named Mario Molina went to work with chemist Sherwood Rowland. Together they set to work to try to figure out if the chlorofluorocarbons would break down in the global stratosphere. Because if they were not being destroyed near the ground—and were inert enough to be observed over the remotest part of the world in the vast south Atlantic Ocean—then what on Earth could possibly destroy them, and what would happen then? In 1974, they published a blockbuster scientific paper in the prestigious journal *Nature* providing the answers.

Molina and Rowland reviewed published measurements by others of the reactions of these gases in the laboratory and realized that more information was needed. They made their own new measurements on how sunlight at different wavelengths might break the chemical down to make chlorine atoms in Rowland's laboratory. They concluded that there would be no destruction at all in the lower atmosphere, and that the CFCs would rise up into the stratosphere intact, where they would finally react with the energetic ultraviolet light from the Sun available in large amounts at higher altitudes. These reactions meant that the CFCs would persist in the atmosphere for some 40 to 150 years once released.[11]

Like the persistent pesticide DDT, or the carbon dioxide from the burning of fossil fuels, human emissions of CFCs have been building up every year since their adoption. We release the CFCs far faster than the Earth can clean them up because there is no significant destruction at low altitudes. When these chemicals reach the upper atmosphere, a high-energy blast from the Sun slowly breaks a CFC up to make chlorine atoms. In contrast to the inert CFC parent molecule, the chlorine atom is a very nasty and voracious offspring that rapidly reacts with ozone, the first step in ozone destruction. Molina and Rowland also sketched out a new catalytic cycle of ozone loss. The catalytic nature of the cycle meant that chlorine atoms regenerated as they ate up ozone. It meant that every chlorine atom could react with ozone over and over again, many thousands of times while it traveled through the stratosphere, destroying ozone all the while. And the long lifetimes of the CFCs meant that they would keep damaging the ozone layer for many decades, even if emissions stopped.

Persistence is a property that should arouse uneasiness every time we encounter it in human impacts on our environment. We fear long-lived nuclear waste because we recognize that nuclear material is dangerous, so the idea that it hangs around essentially forever scares us. In a similar way, a large amount of the carbon dioxide from the coal, methane gas, and oil we burn is also going to be with us for thousands of years. If we're going to do something irreversible on timescales of many decades (or more), we had better make doubly sure what we're doing is safe. And we were doing absolutely nothing about the danger of the increasing amounts of persistent CFCs in the atmosphere.

Rowland and Molina were chemists, and they were onto something that would change the understanding and politics of the ozone layer. Observations of the day didn't suggest that the CFCs were already altering the ozone layer. But Molina and Rowland roughly estimated that the CFCs could destroy as much ozone by about the middle of the twenty-first century as the supersonic plane would within a few decades. They left it to others to do more detailed calculations of how much of the Earth's protective ozone shield would be lost and when. Crutzen was among those who promptly built the mathematical tools to figure that out. Within a few months, he published a paper stating that if the 1972 global usage rate of CFCs were to double, about 10% of the global ozone layer would be destroyed. The production of CFCs was going up at a rapid clip of about 7% per year. At this rate, in only a decade, the planet would see double the CFCs.

This man-made threat to the ozone layer wasn't in the form of a plywood airplane model on a Boeing executive's desk. It was mainly already sitting in medicine cabinets—in cans of underarm deodorant and the hair spray lavished on those curly pompadours and bouffant styles. The tasteful woman of the day wouldn't even go fishing without it. These everyday items were the dominant sources of CFCs in the mid-1970s.

Nearly all of the companies that made spray cans and the chemical industry that produced the chlorofluorocarbon propellant were incensed when they heard scientists saying that their marvelous product posed a risk to the environment. Spray cans were then a tremendous money-maker, with sales growing from about 5 million cans a year in the United States in the late 1940s to 500 million by the end of the fabulous 50s. In 1973, just before the scientific story hit, 2.9 billion cans were sold in the United States. CFCs were cheap, which made them profitable as

New Spray-Set
by the makers of
Lustre-Creme...

SETS HAIR TO STAY
THE SOFTEST WAY!

IDA LUPINO, lovely star of
television program, "MR. ADAMS AND EVE."

SUPER-SOFT

LUSTRE-NET HAIR SPRAY

Loved by Hollywood Stars
because it's <u>non-drying</u>...
contains <u>no lacquer</u>...
mists hair with <u>Lanolin</u>!

Hollywood found it
first...now
it can be yours!
There are 2 types of Lustre-Net.
Super-soft for loose, casual hair-
do's. Regular for hard-to-manage
hair. 5½ oz. can — *a full ounce
more!* Only $1.25 plus tax.

Hairspray advertisement from the 1950s promoted frequent use, even for rec-
reational activities.

a propellant, but other ways to get material out of a can already existed—and more would soon be found. Nevertheless, industry tends to react like cats: their first impulse is that change is simply bad. The spray can business in the United States of the 1970s boasted that its total value stood at about $3 billion, while refrigeration and air conditioning, albeit smaller fractions of the total use of the chemicals, topped out at about $5.5 billion at that time. Industry pushback against the ozone theory included promoting the view that overall the CFCs accounted for anywhere from 200,000 to 1 million American jobs.[12]

If you're skeptical reading that number, so were others at the time. The chemical companies only employed about two thousand people in making chemicals and selling these products. It became clear that the industry was listing anyone anywhere who had any contact with anything with a molecule of CFC in it. As a critic in *Rolling Stone* magazine put it, "They must be down to counting air-conditioner repairmen."[13] Those jobs wouldn't vanish if some other chemical replaced CFCs, although they might change.

CFC production was a fairly big industry, but it was neither as massive nor as organized as the fossil fuel producers are in climate change discussions. More important, the CFCs never became the emotionally charged issue that climate change became, because everyone knew that substitutes were practical in many applications. One of the environmental negotiators from Saudi Arabia once told me in a UN negotiating meeting on chlorofluorocarbons that "We really don't care what molecules are in air conditioning, as long as it cools." A switch from chlorofluorocarbon propellants would hurt some companies but benefit others—the growing thundercloud over the CFCs would boost the market for pump hair sprays and stick deodorants. By the mid-1970s, the Mennen corporation switched their advertising to "Get off your can. Get on the stick." After all, you'd get more

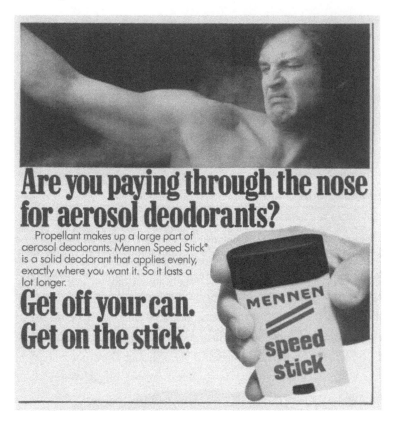

Are you paying through the nose for aerosol deodorants?

Propellant makes up a large part of aerosol deodorants. Mennen Speed Stick® is a solid deodorant that applies evenly, exactly where you want it. So it lasts a lot longer.

Get off your can. Get on the stick.

Mennen corporation advertisement promoting stick rather than spray deodorant, circa 1974.

for your money, with one stick going as far as three spray cans. This slogan skillfully avoided any potential for fractious environmental debate, yet appealed to growing public unease to increase their sales.

The chlorofluorocarbon-producing chemical companies were mostly pretty diverse corporations that made everything from ammonia to xanthan gum (that mystery ingredient often found on food labels). As long as CFCs continued to yield a tidy profit, then producers worldwide would be happy to make and sell them. But these companies typically make vast numbers of different chemicals. At its height, chlorofluorocarbons only represented about 2% of the total business of America's largest maker, the DuPont corporation. The investment was significant, but it wouldn't break the bank if it were lost. Perhaps more important, DuPont and others were always on the lookout to make something else. They are, after all, technical companies, and the challenge of inventing something better is constant and often welcome if the original product isn't a strong seller. In short, substituting something else for CFCs was not as threatening to industry or governments as many made it out to be at the start of what would become known as "the ozone wars." To be sure, there was a pitched battle pitting company against company and government against government (mainly on the two sides of the Atlantic), to sell as much CFC as they could before any phasedown, and to win the contest to dominate a replacement market of substitute chemicals.

For some especially nasty cats, there would indeed be an ozone war. One outraged captain of the spray can industry was the inventor of the leak-proof valves that sat atop those billions of cans in everything from deodorant to bug bombs. He wrote to the chancellor of the University of California to complain that Professor Rowland was too much of an activist, and should be made to spend his time in some other activity. The university staunchly stood by Rowland, but that first experience was a shock to him, as similar experiences were to me and others who picked science for the scholarly life we thought it offered.[14] Johnston had already been in the fray, and cautioned Rowland that industry was only beginning their pushback. "Rowland and Molina were attacked, just like . . . I was attacked," said Johnston.[15] But Rowland was not a person who ever gave in to attempts at intimidation. While the publication of Molina and Rowland's paper was quietly ignored (as scientific papers often are), they jointly gave a press conference at a large scientific meeting to get the word out, and it didn't take long after that for news of their results to get into the papers and on TV.

Scientists and industry were only beginning to fight, but fortunately many Americans proved capable of making up their own minds and taking action. It had only been a few years since the first Earth Day. The environment was very much on the public radar, and skepticism of industry was widespread. By the mid-1970s, one public survey showed that a surprising three-quarters of respondents had heard about the CFC issue, and fully half said they had stopped using aerosol spray cans voluntarily because of their concerns about the environment. Increasing numbers of Americans had simply decided to "get on the stick" to save the ozone layer. It was only a theory; it was far in the future, but large numbers of Americans began stocking the shelves in their homes differently. This simple action thus kickstarted the global phase-out of CFCs and demonstrates the essential role of the public in solving environmental problems.[16]

Americans today can be proud that this happened in our country, because it didn't happen everywhere. The history of sales of CFC shows the spectacular power of US consumers voting with their dollars. CFCs had been a great business for US chemical companies from 1965 through 1974, with metric tons sold increasing year upon year. But by mid-1975 booming business became bust business—not via regulation but largely by vast numbers of people making choices. By 1977, US sales of CFCs had nose-dived by almost 40%.[17] US chemical companies now had a money-losing product that they would be happy to dump in favor of a non-ozone-destroying substitute. Over the years, US industry representatives I have come to know well have told

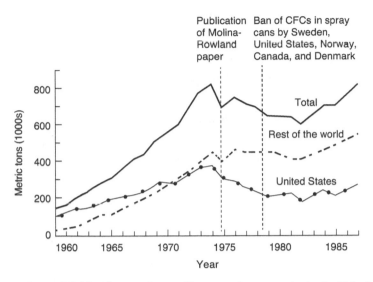

Metric tons of chlorofluorocarbons sold per year by companies in the United States, the rest of the world, and the total, illustrating the decline in the United States even before federal regulation in 1978.

me how important that consumer choice was in forcing their companies to change. In contrast, in other parts of the world, the market for CFCs (dominated by Europe at that time, with a bit in Japan), did flatten but never dropped, so the European companies had incentives to hang on, particularly for lucrative anticipated future sales in the poorer parts of the world.

Every now and then, a powerful industrialist makes an environmental decision centered not on profit but on principles, and it is always inspiring to see. In mid-June of 1975, Sam Johnson made an astonishing announcement about the company that his great-grandfather had founded in 1886. Johnson Wax was then the fifth largest maker of spray can products in America; they sold dozens of household products but are perhaps best known for their Pledge spray furniture polish "to make your home shine." Johnson abruptly announced that his company would be eliminating the CFC propellants from all of their cans. According to his son, this meant a loss of considerable business, as well as attacks from other executives of the spray can industry, who accused him of driving them all to "ruin," despite the fact that other propellants were already available for many things. Johnson's motivation may have also been shrewd self-preservation, since he made it practical for environmentally conscious consumers to keep using his products instead of giving up the sprays altogether. Johnson's well-loved household products would now be emblazoned with a label pronouncing them safe for the ozone layer.[18]

At first, the federal government was slow to respond to the CFC issue, despite widespread public interest. A big part of the problem was that this strangely global issue was so new, and responsibilities were unclear. The Environmental Protection Agency (EPA) had only just been established in 1970, and the CFCs weren't part of its official job yet. The Food and Drug Administration (FDA) regulated medicines, deodorants, and beauty products in cans for their effects on consumers' skin or hair, but the CFCs themselves were safe to handle, even to breathe. The FDA wasn't in charge of indirect effects on the global ozone layer, and neither was anybody else in Washington. So just like lead in paint, pesticides, and other emerging environmental issues, under the US Constitution, the responsibilities initially fell to cities and states.

The state of Oregon was the first to step up, building on broad popular interest that had positioned it as an environmental leader among states for many issues. Oregon had been the first to ban non-refundable beverage containers, for example. Just a year after the Molina and Rowland paper appeared, the state legislature voted to eliminate the use of CFCs in non-essential spray cans, and levied a hefty $1000 fine, or one year in prison, or both, on merchants who violated it. Exempted uses included anything where replacements couldn't yet work as well as the

CFCs (such as asthma inhalers, but these uses were small). Oregon pragmatically gave the sellers two years to get rid of their stock by having the ban take effect in 1977. Soon twenty other states had similar measures in the works.[19]

Within the federal government, it was clear what really needed to be done: both the House and Senate included champions of the environment who set to work to amend the Clean Air Act to confront the problem. The Natural Resources Defense Council (NRDC) played a leading role among environmental advocacy groups, keeping the pressure on them to follow through and legislate. The changes made ended up giving the EPA the authority to regulate CFCs, not just in spray cans but in all applications, should they be proven to be dangerous to people or to the environment. Multiple hearings took place, further informing the public about the issue and helping to define agency responsibilities. An important outcome was a move to put NASA in charge of funding research on the upper atmosphere, which would greatly accelerate the ability of the science community to provide the basis for understanding and action. Researchers picked up their pace and expanded their scope, refining measurement methods that could detect more chemicals in the atmosphere, improving the understanding of chemical reaction rates in laboratory experiments, and developing better computer simulations to make predictions.

While there were many champions in Congress who advanced legislation on CFCs and ozone over the years (including future Vice President Al Gore, when he was still a senator from Tennessee), it was freshman senator Dale Bumpers of Arkansas who took on the matter with gusto. He held a flurry of influential hearings, questioning the industry witnesses with skills he had honed as a trial lawyer. Among other things, he managed to get DuPont's representatives to promise publicly to stop making CFCs at all if "we could look at the data and the data were reasonable and right."[20] After a lot of haggling between different committees, legislation was passed to amend the Clean Air Act in 1976, beginning EPA's role in regulation.

Sweden was the first country to enact a national ban on non-essential CFCs in spray cans, in January of 1978. Sweden had no domestic chemical companies making CFCs, nor did Norway, which followed the Swedes in banning the gases. For these countries, banning CFCs from non-essential spray cans did no harm to their economy—on the contrary, it would even reduce costly imports, since they had to buy these materials from others. But there was not yet a European Union (EU), and the Netherlands, United Kingdom, France, and West Germany were major producers of CFCs, so the European bans stopped short with the Nordic countries for a number of years. The United States did have its own domestic producers, but it also had an engaged public that wanted these chemicals gone. The United States announced a ban on non-essential CFCs in sprays in March of 1978. Canada banned them as well. American CFCs were phased out of almost all spray cans, with the continuing and pragmatic exception of medical applications, especially asthma inhalers. Their use in making some kinds of foams also slowed. Continuing use of CFCs in applications where practical substitutes weren't deemed available, especially in refrigerators, air conditioners, and solvents, were, however, exempted. One environmental campaigner lamented that "we had won only one part of this battle." The exempted uses of CFCs were still growing, along with other very potent ozone destroyers called halons (which contain bromine), which were widely used in fire extinguishers.[21]

It was not until an International Geophysical Year was organized in 1957 that dozens of ozone-measuring stations were established to span the entire globe, an innovation that would prove critical for the ozone issue. One of those ozone-measuring locations was at Halley, Antarctica, where the ozone amounts would suddenly start dropping in the early 1980s, signaling the opening of the ozone hole after twenty-five years of stable data.

There was not yet a measurable downward ozone trend anywhere on Earth in the late 1970s. Just as some places have rainy springs (e.g., Milwaukee), while others are dry, the observed ozone amounts at all of those stations wobbled up and down from day to day or month to month. The reason for the ups and downs is essentially changing weather patterns that blow the ozone around. Early scientific interest in ozone stemmed not from concerns about depletion but about its potential use as a tracer of winds in order to improve weather forecasting. Since scientists were only expecting a small percentage change in a century, this was no surprise. The wake-up call from Antarctica was still a few years away.

But the variability meant that industry could point to the lack of an observed loss of ozone as evidence that no further regulations were required. They argued that further controls were unneeded now that the non-essential use in spray cans had been stopped by regulation in several countries, and overall US sales of CFCs were not growing as rapidly as they had been before 1975. In addition, US public interest in environmental issues and momentum for addressing them declined in the late 1970s. Then Ronald Reagan became president of the US in 1980, ushering in a growing antigovernment and antiregulation movement that initially brought any hope of further prompt action on CFCs in other applications to a screeching halt.[22] The Reagan revolution was the beginning of the antiregulation movement and culture wars that may seem like impossible roadblocks to environmental action today. But concern over ozone depletion would shortly prove that these roadblocks could be overcome.

US industry strongly argued that an international agreement was required for a "level playing field" if any further regulations were to be imposed. Only an international agreement would do to their competitors in Europe what the US consumer had already done to them: reduce their lucrative markets in CFCs for spray cans. If European consumers had done as Americans did, the history would likely have been different. But the conflict over ozone protection in the supersonic transit controversy was still vivid in popular memory, and it cast a long dark shadow of cynicism on European public opinion about ozone depletion.

Science followed a similar transatlantic divide, with the US National Academy of Sciences coming out with a report in late 1979 that was widely viewed as arguing for further CFC regulations, while the UK's Stratospheric Research Advisory Committee followed a few months later and concluded that the entire issue of ozone depletion was in doubt because of inaccuracies in computer models.[23] Just as in other issues, differences of views among credible scientific experts stymied environmental protection and buoyed companies who were seeking to fend off change.

Any further regulations in the United States would not be targeting easily attacked products like spray deodorants that could simply be replaced with roll-ons, but uses that were important not only for human comfort but even for health, most notably refrigeration and air conditioning. It was easy for industry to emphasize how important these applications (and by extension the CFCs in them) were, versus the ever-present uncertainties in the science. The weaponization of uncertainty became enshrined in industry and government tactics, and remains a common refrain in modern environmental discussions.

With no atmospheric measurements to show that stratospheric ozone was already decreasing and predictions of the future change under attack, in 1981 the DuPont company ended its

$21-million research program searching for safer substitutes for the CFCs. As Joe Steed of DuPont put it, "There wasn't scientific or economic justification to proceed."[24]

Anne Gorsuch was Reagan's first EPA administrator. By December of 1981, she was hard at work cutting funding and staff enough to "virtually dismantle the agency" according to one observer. Over at the US Department of the Interior, Secretary James Watt embarked on a similar campaign to slash and burn the agency responsible for forestry and management of public lands. While many religious people view protecting Earth as part of human responsibility toward God's gifts of beauty, Watt is famous for stating that because the Second Coming was at hand, there was little need to protect the environment. Meanwhile, Gorsuch was looking for ways to dismantle the Clean Air Act because she thought smog reductions were already good enough, and the costs of further controlling air pollution were too high. The president of the National Coal Association declared at the beginning of Reagan's tenure that "We're deliriously happy."[25]

It was a grim era for people interested in environmental protection. It's too easy to remember the serious threats and setbacks, and forget the ultimate outcome. The dark time did come to an end. Environmental protection rebounded. Smog controls got better and better despite Gorsuch's opposition, and public lands still grace our American West. Many of President Donald Trump's appointees at EPA, Interior, and other agencies seemed eerily like those chosen by Reagan, and their imprints on our planet, while significant, can fade with time just as those of Reagan did.

Gorsuch resigned in March of 1983, followed by Watt in September of the same year. Watt met a spectacular political end of his own making, when the nation decided that this strange utterance (about the Interior department's coal advisory commission) was just too much to bear from a public official: "We have every kind of mixture you can have. I have a black. I have a woman, two Jews, and a cripple. And we have talent."[26]

While the headlines were focused on the shenanigans of top officials, professional civil service staff at the EPA and the State Department were diligently shrugging off the rhetoric and circumventing their leaders by working for change at the international level. They were part of a working group under the United Nations Environment Programme, along with their hardworking counterparts in other nations. The working group began meeting in 1981 and would convene repeatedly in coming years, starting the challenging process of negotiating the formation of a UN convention and protocol for the protection of the ozone layer. In just a few years, the US would end up playing a leading role in international regulation of the CFCs, putting the legacy of Gorsuch, Watt, and Reagan squarely into the rear-view mirror, thanks in part to the quiet preparations by these staff members.

Just when it seemed that things couldn't get any worse, the US environmental community witnessed what seemed like a miracle. Reagan decided that the best way to restore his administration's tattered environmental image after Gorsuch and Watt was to bring back the respected Bill Ruckleshaus to head up the EPA again in mid-1983. But hopes were quickly dashed when the US National Academy of Sciences released an updated ozone report in 1984 that projected only a 2%–4% depletion of the ozone layer by the late twenty-first century.[27] While scientists stressed that these changes still posed a continuing threat, the industry was overjoyed. Better laboratory measurements of the speeds of some key chemical reactions along with consideration of concurrent trends in carbon dioxide, methane, and nitrous oxide, had made the story less threatening, which always makes communication with the public and policymakers harder. In an attempt to make the newly installed Ruckelshaus do something, the NRDC filed a lawsuit to force the EPA to place a cap on US production of CFCs because of its legal obligations under the Clean Air Act Amendment of 1977 to regulate CFCs if they were harmful to human health and the environment.

Meanwhile in the United Nations, negotiations proceeded apace in preparation for a major meeting in Vienna in March of 1985. Norway, Finland, and Sweden proposed a worldwide ban on the non-essential use of CFCs in spray cans, along with some controls on all uses—an idea called the Nordic annex. The predecessor of the European Union, the European Community, had some big chemical companies of the Netherlands, United Kingdom, and France to protect, so it was adamantly opposed. With NRDC's lawsuit in the background and Ruckelshaus back at the helm, the US antiregulation cohort pivoted on its earlier foot-dragging and embraced the Nordic plan, because it would "level the playing field," as our own industry wanted. The United States formed a bloc that would become known as the Toronto group, including Canada, Sweden, and Norway (and later Japan). The Vienna meeting began looking like it would be quite a mash-up, with the United States in one pro-environmental protection corner and the biggest and most powerful European countries in the other. And then the negotiators did what they do best: they decided to think long term but force something to happen in the short term too. The pro-environment side agreed to drop a threatened demand for a vote on the floor at the Vienna meeting to ban non-essential spray can use of CFCs worldwide. That vote would have embarrassed the Europeans by forcing them to vote no in public to something so obviously practical. Other countries had banned CFCs in non-essential spray cans years earlier and by then, the alternatives were plentiful and cheap, so there was no logical reason for Europe not to do it too. But about 30% of their production was still for spray can use, (including exports to the growing markets in lower-income countries), and that justification seemed quite enough to the French and British governments.

In return for dropping the contentious floor vote, all nations agreed to formally establish in this meeting the Vienna Convention for the Protection of the Ozone Layer, which would negotiate the establishment of an official UN Protocol to reduce production of CFCs at a later date. The Vienna meeting's outcome was only a first but diplomatically significant step. By starting a UN convention, all countries had officially recognized that they had a problem requiring them to work toward a solution. The wheels of diplomacy had begun turning.[28]

As the diplomats negotiated and signed the Vienna Convention, I was a young scientist working on those infamous computer models of the stratosphere, having graduated with my PhD from Berkeley in 1981 and gone to work for the US National Oceanic and Atmospheric Administration (NOAA). I had also become an adjunct professor at the University of Colorado and enjoyed working with graduate students. I was very lucky to be able to collaborate closely with Rolando Garcia, an expert in atmospheric motions and dynamics at the National Center for Atmospheric Research. Together we had been building a computer simulation model that included both chemistry and dynamics of the global stratosphere in a new way that was widely acknowledged at the time as being the best practical approach within the limitations of computers of the time. There were about half a dozen research groups using similar stratospheric models, and we frequently had meetings to discuss our results, many of them organized under the banner of NASA. NASA was taking on more and more responsibility for national and international science of the ozone layer. Amazingly smart and energetic chemist Robert

("Bob") Watson, now known to his colleagues as Sir Bob after being knighted by Queen Elizabeth II in 2012, was heading up NASA's work.

The discovery of the ozone hole using Antarctic station data caused a shock wave through the scientific community worldwide when it was made public in the journal *Nature* in May of 1985. Many of my more senior colleagues were aghast that it had been published. If this was real, why hadn't the satellites picked it up? Solving that mystery didn't take long. NASA scientists went back and rechecked their algorithms and data and reported within a few months that they saw the Antarctic ozone changes too, and they were covering most of the continent. In the view from space, you could easily see that a massive springtime ozone hole had formed. The satellite data could even be made into dramatic video clips, in which you could see a swirling vortex with its ozone hole over the South Pole, rotating around like a hurricane viewed from space, a hurricane with a gaping hole instead of an eye. As these images hit the TV and newspapers, the issue came alive for the public. If it proved to be due to CFCs, then it would be a scientific crisis that was perceptible to people. They had already learned about the threat of skin cancer and other damage from ozone loss when the supersonic transports were making headlines, so they realized that if anything like this ever occurred over their own heads, it would be deeply personal.

The ozone hole had the allure of scientific mystery. The best experts could not be sure why it was happening, but the evidence was clear: it was real. It was many times worse than anyone's model predictions for CFCs. We needed more data if this mystery was going to be solved. The process of getting that data would also capture the public imagination, because Antarctica exemplifies the great unknown and great untamed, a place where scientific heroes go to extremes for the sake of knowledge. This

Sep 15 1999

The ozone hole as observed by NASA satellites on September 15, 1999.

problem wasn't something that one or two scientists would take on as individuals. It was going to take multiple measurements done by teams, as well as other teams working in the laboratory or on increasingly sophisticated models. And the world would be watching, so we needed to do it fast and we needed to do it well. Yikes. But from a scientific point of view, it was the pinnacle of excitement and fun. There is nothing more fascinating to me than attacking a scientific mystery, and what could be more fantastic than one involving the coldest, remotest place on Earth?

Mack McFarland left government scientific research on the ozone layer to go to work for the DuPont company in 1983. He was the key science advisor on the ozone depletion issue until his retirement in 2015. Mack was a very insightful scientist, as well as an honest man, and in his role at DuPont, he became the foremost link between the science community and US industry.

DuPont also had a group of several excellent scientists doing the same kind of numerical simulations of stratospheric ozone and its chemistry that I was doing in the mid-1980s, and they were links too. The DuPont modeling team attended scientific meetings, showed their results, and looked at ours and those of the other modeling groups (which came from several countries, including the United Kingdom, Belgium, and Italy).

These DuPont men (and they were all men) were also highly respected by the scientific community because they were smart and were doing science fairly, according to the high standards of honesty to which all good scientists adhere. Along with Mack, they were constantly communicating with the boardroom at Du-Pont, and the meetings of the Chemical Manufacturers Association in what I am confident was an entirely scientific fashion. Mack described those meetings to me, and he was such a straight shooter that I trusted those accounts.

In the fall of 1985, I started pushing on the model that Rolando and I had developed to try to get it to produce an Antarctic ozone hole. I tried all kinds of chemical ideas, and I considered the constraints. The ozone hole was there in October and maybe September but not in January or May, and we had almost no data about what was happening in the long dark Antarctic winter of June–August, because most of the ways ozone is measured need sunlight. We also knew that there was no corresponding annual Arctic hole. Although there might be some loss in the Arctic, Antarctica was unique.

Antarctica is the coldest place in the world, not just at ground level but also up in the stratosphere. The stratosphere is normally too dry for cloud particles to form, but under the extreme cold of Antarctica, what are known as polar stratospheric clouds can form. Their existence was known not only from recent satellite data, but also from visual observations dating back to the early explorers of Antarctica. But they were considered a curiosity—not a player in the chemical reactions of the atmosphere. Gaseous chlorofluorocarbons break down in the high-intensity light of the stratosphere and are converted mainly to hydrochloric acid and chlorine nitrate gas. As long as the chlorine stays there, ozone is safe. And these two things don't react together at all in the gas phase. But I began to think about what could happen on the surfaces of those icy cold clouds. Surfaces can change chemistry profoundly; that's why our gasoline-powered cars have catalytic converters whose surfaces transform our auto exhaust into less dangerous compounds. I reasoned that hydrochloric acid and chlorine nitrate could come together and react on those polar cloud particle surfaces, liberating ozone-destroying chlorine back to the gas phase. That, along with the need for a little bit of sunlight (to drive additional reactions) was the essence of my idea. I estimated how fast the surface reaction might possibly be if it were very efficient, and just put it in the model. And the calculated amount of ozone in the model went down markedly. Many of my colleagues would view our study as an outrageous idea, because nearly everyone believed the then-conventional wisdom that only reactions between gas molecules could matter in the stratosphere. A few colleagues delighted in telling me that this surface chemistry couldn't possibly work, with one saying "these [cloud] particles don't behave like swimming pools" where reactions could happen. But in the end, it turned out that's exactly how many of the particles can behave. Molina and Rowland's work, Crutzen's work, Johnston's work, everyone's work had all involved gas molecule reactions only, and the chemistry of surfaces was why science had missed predicting the Antarctic ozone hole before it happened.[29]

It was not the whole story by any means, but surface chlorine chemistry on these clouds would turn out to be the key first step in producing the ozone hole, and remains essential to modeling the stratosphere today. The reason why an ozone hole wasn't forming in the Arctic is simply because it is generally much warmer than the Antarctic, although there have been a couple of extremely cold Arctic years recently when depletion there rivals that in the Antarctic.

I presented the work to colleagues at a scientific meeting in that same month. I remember quite well the skepticism with which my presentation was met, with lots of hard questions. But hard questions are what science is all about, and they make our work stronger. It needs to be strong if people are going to make industrial decisions or base policies on it, and I didn't mind. At that time there were three different theories, all trying to explain the ozone hole. Ours involved CFCs and surface chemistry, another involved unusual stratospheric wind patterns, and a third involved nitrogen chemistry from solar activity. The only way we were going to begin to tell which was right was to get more data, not just on ozone but also on the things that affect it, including chlorine and nitrogen compounds, and on chemicals that can track wind patterns. That meeting was the occasion where the scientific community began making plans to go down to the Antarctic as soon as possible and take more data using a variety of instruments.

That's how I found myself on that ice-cold airplane headed to McMurdo Station, Antarctica, my first of several trips there to take data. Doing science in the Antarctic was simply thrilling in every conceivable way: the journey, the stark, pristine, and frigid landscape, topped off by taking data while standing out on a rooftop in winds of 40 miles an hour and air temperatures of –40°F (also –40°C, the only temperature at which the centigrade and Fahrenheit scales show the same value).

There were many long days and nights of taking measurements in the Antarctic when I experienced the thrills of scientific adventure. We took chances we shouldn't have taken, like the night when I was collecting data using light from the moon with a mirror system by myself, and an unexpected windstorm swept in, howling with fury. We normally took down our rooftop mirror equipment when we finished working, to protect it from just such an event but this one happened too fast. There was a good chance that our precious stuff might get blown down, busting our hope of a complete dataset to smithereens along with the mirrors. There was only one thing to do, so I did it: I climbed up to the roof. It was normally easy in good weather, but the wind that night tugged at me ferociously. Just when I grabbed the equipment, a gust pushed me so hard that I almost fell off. I immediately dropped into a crouch to stabilize myself, then slowly crawled back to the ladder on all fours and descended the fifteen feet to the ground, cradling my scientific treasure in one arm.

We measured nitrogen compounds with two different methods and found them to be low, far too low for the solar nitrogen chemistry theory to be right. And we measured ozone, watching in morbid fascination as it decreased dramatically between late August when we arrived and October. So we pinned down the seasonal nature of the depletion, caused by the overlap between extreme cold temperatures and the return of sunlight to the dark pole. Most important, we measured two different chlorine-eating compounds whose concentrations, much higher than in other places, pointed toward the CFCs as the most likely cause of the hole. We had two independent measurements of Antarctic chlorine chemistry, and it looked like a different planet from the one we had been expecting for years.

Perhaps most dramatic were the balloon measurements. Our team from the University of Wyoming did those, hanging ozone-measuring instruments from a big balloon full of helium that lifted them right up into the stratosphere, measuring ozone as they rose. I had the honor of holding one of the ropes a few times, part of a team of four who loaded up the balloon, carefully positioned it to keep it from bumping into the nearby buildings and hills, and then let it go. Most of the launches were calm and routine, but that unpredictable wind was our enemy here too, grabbing the balloon and nearly tossing it to the ground one day when I was helping out. Fortunately, the head of the balloon group was a scientist with many decades of experience in just such situations, and he barked out the command to reel it in like a master sergeant. We stood there helpless, hanging on for dear life as the balloon was buffeted and tossed. I was skeptical when he called "OK, now let go, right now!" but of course he was right, and the balloon floated off exactly where it needed to go, obediently transmitting data back to our receiver on the ground. The balloons demonstrated a very important point: the ozone was missing from a particular range of altitudes, eaten away precisely where the polar stratospheric clouds occur, and not above nor below. The evidence mounted that the polar stratospheric cloud chemistry was indeed at work, and in the ballpark of what was needed to explain the hole.

High-altitude planes flew out of Chile the next year, in 1987. The expedition to Chile, organized mainly by Sir Bob, used NASA's fleet of former U2 spy planes, and included a team from Harvard University led by Jim Anderson. Using yet a third independent method, they too found incredibly high amounts of

Recovery of a balloon payload on snow-covered permanent sea ice by two scientists during the National Ozone Expedition, 1986.

chlorine monoxide in the Antarctic. The clear "smoking gun" was that the amounts of this key chemical were much lower outside the Antarctic over Chile and rose dramatically just where the ozone concentration started falling as the plane flew south.[30] By late 1987, I think it's fair to say that enough of the key results were in and the expert community was pretty well convinced: CFCs caused the hole. I remember Mack questioning my results at a scientific meeting sometime in 1987, and although the questions were tough ones about the instruments we had used, the uncertainties of the data, and potential errors, they were completely fair. It would take a few years for all the data to be scrutinized, peer-reviewed, published, and understood by the full community of scientists involved in the issue, but that is a very short time for something so important to science and society.

In September of 1986, the US industry's lobbying and organizing group, the Alliance for Responsible CFC policy, unexpectedly announced their support for international regulation of CFCs. Although they did not acknowledge that the ozone hole might be due to CFCs, they did say that the scientific information was sufficient to demonstrate that continuing large emissions of CFCs would be dangerous for future generations. They called for a "reasonable global limit on the future growth of . . . CFC production."[31] European industry was furious, and the transatlantic divide on ozone policy became even deeper and more contentious. An urban myth began that is still promoted even by some serious scholars, because it sounds so pat and plausible: that the US companies, including DuPont as the largest and most powerful, must already have developed the perfect substitute chemical and therefore wanted to push for regulation to create their market. But DuPont had stopped its research into alternatives years before, in 1981. According to Mack, that work was far from a solution in 1986, and DuPont had no substitute then, and I believe him.

The US chief diplomat who negotiated the Montreal Protocol was a graceful and intelligent career State Department employee named Richard Benedick. Benedick wrote that there were no secret substitutes on the US companies' shelves. The Alliance's 1986 announcement must have made it more difficult for Benedick to deal with the diplomatic nitty-gritty of moving forward with other governments on the agreement in Vienna to formulate a protocol, although it would be helpful for support within the United States when the time came to act.

Ever since the US spray can ban, DuPont had had more capacity to produce CFCs than the US market could bear. Imagine a half-empty chemical plant somewhere, requiring electricity, heat, and servicing, sitting idle much of the time, and producing CFCs at far less than capacity. That meant DuPont was not only not making money—it was losing money on its investment in the plant. DuPont made a vast range of chemical products, and CFCs were far from their favorite children. Any good company that's not making good money on a particular product will want to move on and make something else if it can, and I suspect that was the quite practical thought not only in the minds of DuPont's executives but also the other US manufacturers. The US consumer's actions had made CFCs what one DuPont executive called a dog of a business.

But DuPont didn't want to spend the money on research (putting themselves at a further disadvantage compared to their European counterparts), unless there would be a market for the substitutes in the things CFCs were still used for. That meant an international agreement was essential to providing the necessary push to spur research and development, such as developing mass production methods for different chemicals that can refrigerate your food instead of CFCs. Without an agreement that forced the chemical companies to develop new substitutes, nothing would happen. After all, the industry's obligation is to maximize profit for shareholders. Policies that are "technology steering" are the vital spurs to environmentally protective innovation in major industries, because without a technology-steering policy there is little incentive to spend money on research.

Policymakers can't push through successful environmental protection policy in a democracy without broad public support. This makes strident public demand for change generally essential to solving environmental problems. Only rarely can the public make big enough changes by their own choices, such as refusing to purchase spray cans. In the case of ozone depletion, the planet was lucky to have that rare case where easy consumer actions could start the ball of change rolling, but technology-steering policies had to happen to take it the rest of the way. The ozone hole provided the next stroke of luck that revved up public interest and support and positioned the policymakers to get something done at the international level.

Benedick worked with staff from NASA, NOAA, the Office of Management and Budget, and the White House's Domestic Policy Council to produce a US negotiating position by Thanksgiving of 1986. The three principles would be (1) a near-term freeze on emissions of the worst chemicals, (2) a long-term, scheduled reduction in emissions, and (3) periodic review. Benedick and his team were eager to work with their counterparts in other governments with similar positions, including Canada, Finland, New Zealand, Norway, Sweden, and Switzerland. They could move on to work jointly on mechanics, targets, and schedules as soon as their carefully crafted US position went through standard approvals. Normally, such approvals are purely mechanical at that point, since the position itself had been developed based on the standard process, with the involved agencies' representatives at the table.

The European CFC companies had little on their minds except preserving the dominance of the world market that they had enjoyed since the United States had banned spray cans. It didn't help that the European Commission (EC) was preoccupied with forging the European Union (which was formalized in 1993) rather than issues like international environmental protection. The European Commission based its positions and tactics in the diplomatic discussions concerning ozone "largely on self-serving data and contentions of a few big companies" according to Benedick. The British and the French were the most reluctant countries to agree during the formulation of the Montreal Protocol, while the United States was almost always the leader among nations calling for a strong protocol with binding targets and timetables. It seems quite hard to imagine those roles, given how different they are in climate change negotiations today, but just knowing they did exist demonstrates that remarkable changes are certainly possible—and could happen again.

Progress on preparing for a meeting to sign a major international agreement went very slowly while the United Kingdom held the position of the rotating presidency of the EC in the second half of 1986. But when Belgium replaced it in January of 1987, the EC stance began to change. Many eyes turned optimistically to the planned September 1987 high-level meeting in Montreal, hoping that a global protocol would be signed.[32]

But in early 1987, a major pothole developed in the US road to Montreal. Several officials in the Reagan administration complained that the points carefully crafted by staff from their own agencies led by Benedick had not been approved at a high enough level. They questioned whether the State Department's position should be tossed out and a new US plan formulated. A series of marathon meetings, scientific briefings, and closed-door fights ensued. Fortunately, one of the foremost detractors of the State Department's plan was Secretary of the Interior Don Hodel, who proved to be no more skillful at keeping the gun off his own foot than James Watt had been.

Hodel proposed that, instead of reducing CFC production, the United States should propose a program of "personal protection" in the form of hats, sunglasses, and tanning lotion. But of course, it's not just people who are threatened by UV light. When the press got wind of his proposal, it came up with one of the most amusing newspaper cartoons in the history of the ozone issue, which portrayed hats and sunglasses on everything from birds to rabbits. Republican Senator John Chaffee, one of the protocol's supporters, scoffed that this approach was as absurd as an earlier Reagan-era gaff of trying to classify ketchup as a vegetable in school lunch programs.

George Schultz at the State Department stood fast, insisting on what he still describes as the need for an "insurance policy" against the risks. Hodel's remark weakened the opposition's efforts, although he protested that he had been misquoted. Only the president could decide whether to change the US position as outlined in the State Department document. In June of 1987 President Reagan sided with Schultz, giving State the nod to go to Montreal to iron out the three points with other countries. Whether the antiregulation champion Reagan did so because of the three skin cancers that had already been removed from his nose remains a matter of speculation. But while sporting a bandage from a recent medical procedure during a public appearance, he did tell the audience that his own nose was a warning to stay out of the sun, indicating that he knew full well the personal connection between UV light and the health of his skin.

International staff met repeatedly in preparation for the high-level meeting in Montreal, under the auspices of the United Nations Environment Programme (UNEP), headed at the time by a smart, skillful Egyptian named Mustofa Tolba. Tolba did a great deal to bring the protocol to fruition, particularly when it came to negotiations between higher- and lower-income nations. He was respected by all and representatives from the lower-income countries trusted him more than they would have trusted anyone who was not from such a country. A cornerstone of the protocol was the concept that would later come to be called "common but differentiated" responsibilities, that the higher-income nations would take on obligations of controls first, while the lower-income nations would have a grace period before any reductions would occur for them. This was only fair, because many of them had so little of the benefits the CFCs offered, such as refrigeration, and because it meant that the wealthier nations would pay

39

any higher costs of the first substitutes, and the lower-income nations would enjoy the economies of scale that would occur as use expanded over time.[33]

Rather than emissions, the Montreal Protocol controls production and consumption of CFCs, because it is relatively easy to keep track of how much is made by chemical companies and sold between countries (exports and imports), and it is much harder to track emissions. Emissions depend, for example, on the standing stock of refrigerators that are leaking at uncertain rates, depending on how old the equipment is and how well it has been maintained. If production and consumption (defined as production plus imports minus exports) were to be frozen, then an agreement would have to be struck about the year for the freeze. The negotiations around that chosen year illustrate the tortuous, sometimes comical and always unimaginably slow way that international agreements are built. I have sat through hours of negotiation in UN working meetings on climate change as well as some on ozone depletion, listening to the diplomats talk around each other incessantly in the most flowery and polite language you can imagine.

It's eye-opening for a scientist to see diplomacy in action, which could not be more different from the normal brass-tacks, get-the-job-done-efficiently scientific discussion. I think this is why nearly all diplomats actually love science. It's a basis for agreement on fact instead of form, truth instead of camouflage. But when to freeze was a policy question and not a science question.

The diplomats from twenty-four countries signed the Protocol on Substances that Deplete the Ozone Layer in Montreal on September 16, 1987. They came mainly from North America and Europe, along with Japan, New Zealand, Australia, and a few lower-income countries. In the words of one representative, "after long and staggering negotiations, strong US pressure had finally forced (Europe) to accept an agreement." At this stage, the few lower-income countries that attended had limited interest in the protocol. These chemicals were part of an alien lifestyle that they could not afford. An Indian representative, for example, felt that ozone depletion was a "rich man's problem, rich man's solution." The involvement of lower-income countries was to change dramatically in just a few years, and would ultimately become the protocol's biggest challenge.[34]

A whirlwind of events took place between 1988 and 1990, in a dizzying sequence that rivaled the furious pace of 1985 to 1987. It involved diverse actors: industry, scientists, governments around the world, and the public. The treaty called for a freeze on production and consumption of CFCs as soon as the treaty "entered into force," followed by a 20% reduction by mid-1994 and a 50% reduction by mid-1999. These targets and dates were subject to revision if the governments chose to do so—the treaty stipulated that the nations (also called the Parties to the Protocol) would meet periodically to revisit the agreement. But the treaty was clearly a strong signal of regulation. Industry worldwide was being sent a loud and clear signal that the CFCs were on their way out. And the treaty also included a measure that may have seemed innocuous but would become a remarkable breakthrough (Article 6), which stated that the parties would evaluate the control measures on the basis of "scientific, environmental, technical, and economic" assessments to be prepared for their negotiations by panels of qualified experts.

Entry into force of UN treaties requires countries to go beyond the easy ceremonial step of signing a pledge to do something and take national actions according to their own laws to commit to it. Mexico was the first country to ratify the Montreal

Protocol. The United States was the second, following a unanimous vote of 83–0 of the US Senate on March 14 of 1988.[35]

It may seem inconceivable today to imagine bipartisan agreement on an international environmental treaty, given the political upheavals of the past thirty some years, let alone unanimous agreement. The ozone hole had alarmed the public so much that all of the senators felt compelled to cast a "yea" vote or join the seventeen who abstained.

Just a day after the Senate vote, the ever-active Sir Bob released a new scientific assessment report that he had chaired, involving ninety-nine other scientists from countries around the world (of which I was one). The two-volume Ozone Trends Panel Report would become known as "the red books." It presented alarming new findings on observed changes in ozone worldwide. Global ozone depletion was not a far-distant risk but was already happening now, right over our heads. The report also presented the striking news that the world's scientists jointly found that "the weight of evidence strongly indicates that manmade chlorine [is] primarily responsible" for the Antarctic ozone hole. Such agreement on an environmental issue among so many experts packed a punch. They also noted that the United States saw a decrease of ozone of up to 6% in the winter, while in South America, Australia, and New Zealand it was much worse. Not only did fragments of the ozone hole sometimes drift over populated areas, but systematic downward trends were also occurring and were larger than those in the North. Both Australia and New Zealand instituted a slip-slop-slap health campaign to combat the potential for skin cancer: "*Slip* on a shirt, *Slop* on the *50+ sunscreen*, *Slap* on a hat."

On March 24, the CEO of DuPont abruptly announced that the company would stop making CFCs altogether, declaring that this dramatic reversal was because of the new science just reported. Mack McFarland had been one of the ninety-nine scientists sworn to secrecy until the release of the report, but he was fully aware of its content and ready to talk to DuPont senior management the instant that occurred. He described the days following the release of that report as a succession of meetings, in which he and his supervisor worked to explain the new findings to others, answering their questions and justifying the key conclusions in detail. Just twenty days before, the DuPont CEO had written to three US senators defending the company's continued production of CFCs; their 25% share of the global market was deemed safe enough in the letter of March 4 because "the scientific evidence does not point to the need for dramatic CFC emission reductions." Sir Bob's red books changed all that, literally in a matter of days.[36]

The chemical industry stepped up work on hydrochlorofluorocarbon (HCFC) and hydrofluorocarbon (HFC) substitutes for CFCs. These compounds are close cousins of the CFCs, with many similar properties when used as coolants in air conditioners and refrigerators, as well as in certain foams and other uses. CFC, HCFC, and HFC may sound like alphabet soup but there's so much magic in those letters. First of all, compounds containing fluorine don't damage the ozone layer at all, in contrast to chlorine. That's because the fluorine ultimately gets neutralized for all practical purposes, tied up in hydrofluoric acid, which is practically inert in the stratosphere. In contrast, the chlorine in hydrochloric acid is much more loosely bound, raring to get out and eat up the ozone layer. Both HCFCs and HFCs also have a fabulous edge on the CFCs when it comes to environmental damage. They have much shorter atmospheric lifetimes than the CFCs, because of the special properties of hydrogen. This means they don't suffer nearly as much from the persistence problem that made the CFCs so damaging. Compounds with a carbon-hydrogen bond can break down in the lower atmosphere, cutting off nearly all the chlorine before it can reach the stratosphere. The shorter lifetimes mean that while both HCFCs and HFCs are strong greenhouse gases, their effect is not as strong as the CFCs they were replacing, a property that would come back into the discussions later, in the twenty-first century.

In 1988, companies in Belgium, West Germany, the Netherlands, and the UK announced that they would voluntarily begin phasing out production CFCs for non-essential spray cans. While DuPont had announced that they would stop producing CFCs altogether (whether for refrigeration, air conditioning, solvents, foams, or any other non-essential use), these European companies had finally decided to do what the United States had done more than ten years earlier, noting that they were in the "fortunate position" of being able to copy the spray can technologies already long in use in the United States to find substitutes.

In September of 1988, British Prime Minister Margaret Thatcher delivered a speech to the Royal Society, the British national academy of sciences. She declared that the United Kingdom would be raising the priority assigned to environmental protection, especially ozone and climate protection.[37] This left only France as the remaining European holdout.

Industry worldwide began to focus on a range of practical alternatives. Many CFCs were commonly used as cleaning solvents. But in short order, Japan developed extremely efficient CFC recovery capability, keeping over 95% of the solvent material contained. And then the news got even better, when it was found that lemon juice could actually clean well enough to replace the CFCs as solvents in many cases. Similarly, many foams could be blown with other compounds, including the HCFCs, and in some cases foam could be replaced with other materials—today's Big Mac comes in a cardboard container, not foam. That happened largely because of consumers who demanded change. Alternatives replaced CFCs in application after application, as the protocol steered technology change. US automakers accepted the use of recycled CFCs for auto air conditioners. And the simple act of using better hoses in auto AC systems (at a grand cost of about $35 per system) led to a large reduction in CFC leaks. Refrigeration and air conditioning turned to the HCFCs and later to the HFCs. The bromine-containing compounds used in fire extinguishing, called halons, would be harder, but some success came early, when it was realized that a great deal of halon emission was caused by practicing to fight fires (especially by the US military) rather than actually fighting fires. Practice sessions switched to using less damaging chemicals. Practical solutions were found time and time again, once the incentive for them was firmly established by the Protocol.

Industry representatives were forced not just to focus on alternatives but to talk to each other about them, because the treaty mandated that reports be prepared not only about the science but also about technology and economics. The Technology and Economic Assessment Panel (TEAP) reports never received the public attention that the science reports did, but they were just as important, because they would bring together the engineers and technicians in industry to share what they knew, which was the only way they could influence the targets and timetables, irrespective of proprietary concerns. Unless they were at the table, changes might be decided upon that they would not be able to implement, making their lot even more difficult. If they participated constructively, they could help shape the next steps, and they did.

More news poured in from the science side in late 1988 and 1989. A field mission to the Arctic found evidence of atmospheric chemistry similar to that happening in the Antarctic. In 1988 Margaret Tolbert and coworkers reported new laboratory measurements of the same reactions that we had identified as responsible for the ozone hole a few years before, but this time the surface reactions were happening on liquid sulfuric acid/water particles at relatively warm temperatures in the laboratory. I immediately knew what I had to do, and immersed myself in all the data that was available after the eruption of the El Chichón volcano, which had dramatically increased the amount of sulfur particles in the stratosphere, and used my numerical model to estimate how much ozone loss these reactions could drive at a latitude of 40°N. Sure enough, the model along with post-Chichón measurements strongly indicated that this new surface chemistry mechanism would make CFCs much more damaging to ozone, especially after certain volcanic eruptions but even in non-volcanic years.[38] One colleague described this as a "a potential bombshell."

Around this time, scientists began to highlight an even bigger problem: the 50% reduction target of the treaty wouldn't even be enough to stabilize concentrations of CFCs. They would keep increasing, because at that level, emissions would still vastly outstrip the volume of chemicals that would decay from the atmosphere. Persistence rears its ugly head again whenever our emissions occur at a rate greater than the rate of decay. Every year's emissions would add to the environmental burden, and the CFCs had long lifetimes of a half a century and more. That meant that a 50% reduction was far from enough even to keep those chemicals at the same level, let alone to start decreasing them. The Montreal Protocol as written would slow the problem's growth, but was nowhere near enough to stop it, and far from what was needed to begin to heal the damage already caused.

The diplomats began meeting in preparation for a June 1990 conference at which the targets and timetables would be revisited for the first time since Montreal. The United Kingdom, freshly converted to activism, announced a pre-meeting position of a 100% phaseout of CFCs by the year 2000. The United States made the same proposal. More than a hundred countries now attended these meetings, along with many nongovernmental organizations, such as Friends of the Earth and Greenpeace, as well as industry associations. There was wide agreement on the science as well as the technological leaps and bounds that had occurred, but the steamroller pushing toward further controls hit an unexpected wall: the lower-income countries had a new negotiating position. They had realized that as they got richer, they were going to need all the many benefits of CFCs that other countries were already enjoying, and they were not prepared to pay more for those things than the wealthier countries had paid. A grace period during which they would be able to keep using and producing CFCs after the wealthier countries stopped would in principle allow time for substitutes to be developed and economies of scale to make them cheaper. But that wasn't enough. What if the substitute HCFCs or HFCs remained more expensive than the CFCs, what then was in store for them?

When the parties sat down to wrestle with the difficult question of aid to lower-income countries, the leadership of the Egyptian Tolba was a major factor that led to success, although the negotiations still weren't easy. After a series of intense meetings, the agreement was that a new, separate multilateral fund would be established within Tolba's UNEP, financed by cost-sharing among the wealthier countries. Importantly, it would cover all the *incremental* costs to lower-income countries of compliance with the protocol. In other words, if China became a party to the protocol and had to buy refrigerators that each cost $50 more than they would have with CFCs, the fund would pay the difference (but only the difference and not the total cost). This innovation made becoming a party to this "rich man's problem" fair for the lower-income countries.

The proposed fund would cost the United States $20 million a year. The United States took longer to agree to pony up than the Japanese and Europeans did, until the protocol drafters came up with a clever way to find the money. Industry was poised to reap stratospheric profits of $2–$7 billion per year as the phasedown of the CFCs in high-income countries drove down supplies and sent the price skyrocketing. Imposing a windfall profits tax on every pound sold made the alternative technologies competitive and stimulated the market for options, while at the same time providing the treasury with far more than needed for the US contribution.

The parties met in London in June of 1990 to consider amending the protocol they had signed only three years before. A sticking point proved to be a push for a fast phaseout by Europe, but the United States wanted to go slower. The United States had a much larger stock of equipment using CFCs than the Europeans—an infrastructure commitment in the form of a lot more auto air conditioners and air-conditioned buildings. Too fast a phaseout could make it hard to let these age out gracefully or be serviced with new coolants using slight modifications. After considerable back-and-forth, they agreed on an 85% reduction by 1997, followed by 100% phaseout by 2000. The additional 15% for three years had almost no effect on the future of the ozone layer because it was so much smaller than the starting point and lasted only a few years, but it gave the American side its needed cushion. It passed because it made sense, as so much in the ozone negotiations did. The Montreal Protocol negotiations often struck me as logical, with defensible choices in the end, rather than suffering from the zealotry that sometimes seems to impede progress during climate change negotiations. I believe that's at least partly because the diplomats could perceive the science underlying the rationale and act on the basis of easily understood evidence and facts, and because the TEAP always did an extraordinary job of showing what would be practical. The parties left London rejoicing at the plan to eliminate CFCs and other ozone-depleting substances in higher-income countries by 2000.[39]

In the end, the extra time wasn't even needed, as the stars aligned for the Montreal Protocol once more when the parties met next in Copenhagen in 1992. The scientific communities on both sides of the Atlantic carried out intensive field campaigns targeting Arctic ozone chemistry, and found plenty of evidence that the Arctic was primed for massive ozone losses similar to those obtained in the Antarctic if the Montreal Protocol didn't protect it. In the winter of 1992, Arctic ozone dipped by up to 20% and for a few days, ozone depletion over Russia reached the astounding value of 40% below normal. Now the issue was striking much closer to home than the remote Antarctic.

The phaseout again proved to be even easier than planned. Industry developed alternative processes and compounds at a rapid clip, and non-CFC-based refrigerators became widespread. After some squabbling about when the substitute HCFCs would also meet their end, the parties agreed to eliminate CFC production in higher-income countries by 1995, with extremely minor and very practical exemptions like medical applications.

History repeated itself just a few years later at a historic meeting in Beijing in 1999, when the lower-income countries also agreed to stop all production globally by the year 2010. This, too, had become practical because by then the substitutes were both readily available and increasingly reasonably priced, and the multilateral fund was demonstrably working.

The global phaseout proved to be remarkably successful, and observed amounts of CFCs slowly began to decline. In 2016, I published a paper in *Science* that is widely acknowledged as the first to show that the Antarctic ozone hole is starting to show signs of healing.[40] Because of the long lifetime of the CFCs in the atmosphere, the hole is not expected to go away until about the middle of the twenty-first century, but it is starting to form later in the spring season, is less deep, and smaller in area on average (although deeper and larger holes will still occur sporadically in unusually cold Antarctic years). There have been a few signs of a small amount of cheating going on, with unexpected emissions that appear to signal rogue production after 2013, mainly from China. While these are important to understand, they appear to have stopped and are too small to significantly upend the progress toward healing our planet that has already happened. Their prompt detection by high-quality measurements shows once again the power of science in supporting environmental policy.[41]

Protecting the ozone layer is rightly called the world's greatest international environmental success story, celebrated globally and held up as evidence of what people can achieve in managing environmental risk. We were helped when solving the ozone problem by an ideal confluence of what I call the "three P's.": it had deeply personal health impacts, its science was readily perceptible to non-experts, and solutions were eminently practical.

Preventing ozone depletion had the tremendous benefit of a powerful kickstart caused by consumer action to turn away from CFCs in spray cans in the United States, a personal choice that destroyed the market for American CFCs and made our producers eager to look at alternatives. It was an easy thing to do, and it made people feel empowered and interested. Can progress still happen when it depends not on personal choices but policies? Sure. Getting rid of CFCs in refrigeration and air conditioning was not something the individual consumer could do, beyond limited personal choices. But technology-steering policies under the protocol inspired the required innovation to find those solutions and make them cheap enough to be practical. The producers, especially the US chemical companies, deserve credit for the exceptional scientists they nurtured and apparently heeded as the issue developed, rather than turning a blind eye. After a faltering start caused in part by smoldering anger about supersonic planes along with the growing pains of forming the European Union, the British and French eventually came around, and their companies also began to see the logical way forward.

The discovery of the Antarctic ozone hole stunned the entire world and made the ozone issue a "hot crisis." People are much better at solving hot crises than they are at dealing with slow ones. The public's fascination kept scientists energized and politicians well motivated to act. Scientists grew accustomed to the awesome power of teamwork: in field experiments to the ends of the Earth and most important, in international assessment reports. We stopped working as lone wolves and became an effective pack, aware that as a group we could serve the world better than any of us could on our own.

As of 2020, the Montreal Protocol stands as the only UN treaty of any kind that has been signed by each and every member nation of the UN. Ozone depletion will go down in history as the clear case where humankind found itself on the brink of disaster and found the will and the way to pull back. But could we do anything halfway as grand on a slowly evolving problem, like climate change?

2

Greenhouse Gases

KIGALI SHOWS SUCCESS
IS STILL POSSIBLE

After a 2018 lecture about ozone depletion and the Montreal Protocol at a public museum in Sweden, one of the first questions came from a woman who, like many of my students, grew up in in the twenty-first century. She was polite, but skeptical. "That is a wonderful story," she said, "but it's from so long ago. People had different attitudes then." She went on to lament the lack of international agreement, the adversarial relationships between many parts of the world, and the plague of misinformation spread especially on the Internet. "Do you really think we can still solve environmental problems like we could in the twentieth century?" she asked.

The Internet is a huge new challenge, a blessing and a demon that has distorted the world of information. I also believe that the international situation has become much more complicated as lower-income nations have—quite rightly—continued to develop and gained more political clout. But there is cause for hope. I point to the European ban on single-use plastic plates, straws, and utensils like spoons. Many places in both Europe and the United States have also banned the bane of many wilderness lovers: single-use plastic bags. Because a lot of the world's people live in rich nations, these decisions drastically reduce the size of the markets to sell these products, spurring the development of biodegradable, plastic-like substances. In the face of this incentive, researchers are making great strides. With these new approaches, and increases in recycling, I look forward to the next decade when single-use plastic products will largely be a relic. "But what about our biggest problem, climate change? All we have is a voluntary Paris Agreement that has no teeth," she laments.

You might share her fears and frustration. I do too. But did you know that the world already has a binding international agreement expected to phase out one significant driver of climate change? The 2016 Kigali amendment to the Montreal Protocol has entered into force. That's UN-speak for "enough countries have ratified it that it's starting to take action." And it's expected to help reduce global warming by 2050. Much more is needed of course, but every tenth of a degree of warming that is avoided saves lives and reduces hardship. More important, it shows that we *can* still agree to do good things about our environment, even when it comes to climate change. Maybe we should be asking, "*Why* exactly did governments agree to Kigali, and can we translate what happened there to other climate drivers like carbon dioxide?"

In 1975, Veerhabhadran "Ram" Ramanathan was a young post-doctoral associate at NASA's research center in Langley, Virginia.

A physicist and expert in how infrared light penetrates the atmosphere, he became intrigued by Molina and Rowland's pioneering paper on chlorofluorocarbons and ozone loss. He decided to have a look at whether these interesting and novel gases could also contribute to the greenhouse effect.[1] Just like people or animals, our planet has a temperature, and every warm body emits infrared light—night-vision goggles let you see animals in the dark using this property. Many atmospheric gases have the potential to absorb the infrared light emitted from the Earth's warm surface, keeping it from escaping to space and thereby ultimately heating the planet. The amount of energy absorbed depends on the particular structure of the molecules floating through our air. The chemical bonds of some atmospheric molecules vibrate like springs and rotate when they absorb certain energies of infrared light. The motions triggered by that energy turn light into the heat that generates the Earth's greenhouse effect. Most of our planet's greenhouse effect comes from the strong absorption of infrared energy by two gases, water vapor and carbon dioxide, as well as clouds.

The most dramatic illustration of the power of a planet's greenhouse effect is the fact that while Mercury is the planet closest to the Sun, it has an average surface temperature of about 185°F, while much more distant Venus has an average temperature of over 800°F. That's because Mercury doesn't have much of an atmosphere, while Venus has a very thick one, chock full of infrared-absorbing carbon dioxide. In fact, carbon dioxide accounts for around 95% of Venus's atmosphere, so its atmospheric molecules are engaged in a frenzied dance of strong infrared absorption and massive heating. Mars, in contrast, is not only farther from the Sun but also has a very limited atmosphere; its average surface temperature is an inhospitable –80°F. Our Earth is the goldilocks planet—just right. Lucky Earth has an average surface temperature of around 50°F, and before humans began to burn coal, natural gas, gasoline, and diesel "fossil fuels," our carbon dioxide was around 270 parts per million by volume (ppmv, many thousands of times less than Venus's). As of 2022 it's nearly 420 ppmv. Measurements show that carbon dioxide increases come mainly from the burning of fossil fuel, which has allowed a comfortable modern lifestyle in wealthier nations since the industrial era, but threatens us with an uncomfortable and even disaster-ridden future. On starry nights one can gaze at Venus and ponder the risks of adding more carbon dioxide to our atmosphere, which will certainly make us hotter.

Ramanathan realized that a very special feature of the chlorofluorocarbons, as well as any other chemicals with carbon-fluorine or carbon-chlorine bonds, is their ability to absorb certain types of infrared energy ignored by carbon dioxide, water vapor, and other gases in the Earth's atmosphere. The CFCs all absorb in a special atmospheric "window," where there's a lot of light available because other atmospheric gases don't significantly absorb there. When Ramanathan's 1975 paper was published in *Science*, the recognition of the potential for anthropogenic climate change was in a relatively early phase, but he alerted the world that continuing emissions of the chlorofluorocarbons might someday pose risks—not only for the world's ozone layer but also for its equitable climate.

Fresh off the success of the Montreal Protocol in phasing out chlorofluorocarbons, the world's diplomats gathered in Rio de Janeiro in 1992 to consider climate change and sustainable development. It was clear that in dealing with climate change, with its links to the use of fossil fuels, they now had the world's biggest tiger by the tail, a problem whose challenges dwarfed all those that came before it. As the country representatives did when they met in Vienna in 1985 to weigh the idea of a treaty on ozone depletion, in Rio they decided only to organize a convention, to foster further discussions and consider next steps. When in doubt about what to do, talk. Thus the United Nations Framework Convention on Climate Change (UNFCCC) was born. Although the diplomats left the gargantuan problem of sustainable development on the table, they did start talking about an international approach to dealing with climate change.

The United States lobbied hard for the idea that trading of gases offered flexibilities and cost savings, reflecting the ongoing tensions between environmental protection, costs, and regulations that continued to rage across the nation. Five years later, after much negotiation, the diplomats met to formalize the 1997 Kyoto Protocol, signed with great enthusiasm by then Vice President Al Gore on behalf of the United States, along with representatives of more than 150 other countries. The United States succeeded in including trading among different global warming gases, as part of the Kyoto Protocol's mechanics: carbon dioxide, methane, nitrous oxide, as well as the hydrofluorocarbons (HFCs) that were rapidly replacing the ozone-damaging chlorofluorocarbons as the Montreal Protocol kicked in. The hydrofluorocarbons absorb in the atmospheric window and do cause climate change. But they don't destroy ozone because they don't contain chlorine, and were therefore not initially included under Montreal. Then their use skyrocketed as CFC substitutes. There was substantial debate about including the chlorofluorocarbons in Kyoto too—after all they are also potent greenhouse gases as Ramanathan had shown—but the logical argument that the ozone-depleting gases were already being successfully controlled under the Montreal Protocol prevailed.[2]

Ironically, the United States never ratified the Kyoto Protocol that it had helped to craft, and ultimately withdrew from it altogether. Climate change is sometimes called the "mother of all environmental problems," a subject that sometimes seems so intractable as to induce utter despair about our planet's future. But we *did* set the HFCs on a clear phaseout path, and we did that just a few years ago, in the twenty-first century, not that bygone era of the last century. So I think the right question to ask is, Why did that happen, and what's different about *other* greenhouse gases, especially carbon dioxide and methane?

A comfortable, twenty-first-century lifestyle stems from some conveniences that have become so routine that they often go unnoticed. In the United States, refrigeration is largely taken for granted, but is essential not only for food storage but also for some medical products and procedures. Air conditioning allows cities in hot climates to flourish. HFC coolants have helped this technology thrive over the past twenty years.

For centuries before HFCs, refrigeration relied on the simplest of methods: ice was chopped out of rivers and lakes, covered with sawdust for insulation, stored in icehouses, and delivered to businesses and homes in massive chunks. As early as the nineteenth century, technology for cooling with ice advanced when engineers found ways to make ice and keep it cold, rather than just using what nature supplies.

The fundamental principle underlying nearly all refrigeration and air conditioning is a common experience on very hot summer days, especially in a dry climate: the water of sweat turns to vapor by wicking away heat from your skin. That conversion from liquid to vapor is why sweating cools you off, and the same principle is the basis for modern cooling systems. Refrigeration and air conditioning involve compressing a gas to form a liquid by mechanical means, then evaporating it to generate cooling by reducing pressure so that it boils. Like the evaporation of sweat on your skin, that's a cooling process. This is followed by

Refrigeration in the World War I era. Heavy work formerly done by men only is being done by girls. The ice girls are delivering ice on a route, and their work requires brawn as well as the partriotic ambition to help.

compression to liquefy the gas and use it over and over. Some gases are much better at cooling than others, depending upon how easily they liquefy and boil, and how much heat exchange is involved in those processes. Most early American refrigeration used ammonia, a gas that could theoretically take up more heat than other refrigerants, although its technology did suffer from inefficiencies. But a much worse problem is that ammonia is highly toxic and hence extremely dangerous if it leaks out of a refrigeration machine. It also carries another, less often recognized risk: ammonia can trigger explosions when mixed with air. An explosion and fire at the Chicago's World's Fair on the afternoon of July 10, 1893, was a spectacular object lesson about this risk. It killed seventeen people, including nine firemen, many of whom had to leap from a burning tower to their deaths. At the time, the connection between ammonia and explosions was not understood, but as more and more accidents occurred, it became clear that the benefits of ammonia refrigeration did not come without substantial risks.[3]

Refrigeration using carbon dioxide began to replace ammonia just a few years later, mainly in Europe. Carbon dioxide is not toxic or flammable, but liquefying it requires very high pressure, and that creates a different risk. The engineering challenges of carbon dioxide refrigeration took time to solve. During that time, ammonia technology spread throughout the American industrial market. On top of that, carbon dioxide doesn't cool efficiently in hot climates like those found in much of the United States. Carbon dioxide–based refrigeration therefore has never caught on in the United States as it did across the Atlantic.[4]

By the 1920s, the cooling industry yearned to expand into the consumer market and provide large numbers of home refrigerators, but safety concerns stood in the way of widespread adoption. Appliance manufacturer Kelvinator produced one of the first home refrigerators, using sulfur dioxide. Although quite toxic, sulfur dioxide at least avoided the flammability problem. Tensions grew with consumer demand and grave safety concerns. When the CFCs were introduced by the chemical industry as a potential refrigerant in the 1930s, they seemed like a miracle. Non-flammable, with excellent coolant properties, high efficiency, and no toxicity, CFCs became the kings of global refrigeration and air conditioning until concerns about ozone depletion and the advent of the Montreal Protocol ended their reign in the late 1980s and 1990s. As the phaseout of the CFCs advanced, substitution for them in refrigeration and air conditioning increasingly relied on the HFCs, especially one gas called HFC-134a. Although this gas and other HFCs cause no significant ozone loss, it is a potent "super-greenhouse" gas: emitting a pound of this gas causes over a thousand times as much warming in the next hundred years as a pound of carbon dioxide. This ratio is called the global warming potential, or GWP, in technical jargon, but here I'll simply refer to it as the relative impact on climate per kilogram emitted. The potency depends not only on the strength of the absorption of infrared energy per molecule of a chemical but also on the chemical's time of persistence in the atmosphere, its lifetime. In general, the HFCs were much more "climate-friendly" than the CFC-12 that they were replacing, because CFC-12 lives for about a hundred years in the atmosphere, while the HFCs live a few years to a couple of decades. Carbon dioxide is the dominant greenhouse gas in the world today because we emit many more tons of it than any other greenhouse gas—but that doesn't make those other gases negligible. And as the Kyoto Protocol began to take effect in the 2000s, governments and industry recognized that it was time for a new generation of even shorter-lived and therefore more climate-friendly CFC substitutes.

HFC-134a became the coolant of choice for auto air conditioners of the 1990s in both the United States and Europe. But as one of their contributions to reducing greenhouse gas emissions under the Kyoto Protocol, the European Parliament initiated a phaseout of the HFCs used as coolants in auto air conditioning. The EU passed a regulation in 2006 that would reduce the use of HFC-134a, beginning with 2011 auto models. The European legislation was cleverly designed, even diabolical from the viewpoint of the American chemical industry. It made it illegal not only to use HFC-134a but even to use any replacement refrigerants with a relative greenhouse effect more than 150 times that of carbon dioxide in an auto air conditioner sold in Europe.[5] The limit of

150 strategically aced out another leading candidate chemical substitute then under development for potential future use in the United States (which had an estimated relative greenhouse effect very near 150). This policy ignited what is sometimes called a "cold war" within the industry.

The European law was a form of technology steering, designed to spur the use of carbon dioxide in auto air conditioning systems. The auto industry worldwide would have to fall in line, like it or not, if they wanted to sell cars to Europeans. It may seem surprising that carbon dioxide is an environmentally friendly choice for refrigeration and air conditioning. But in contrast to the carbon dioxide released when fossil fuels burn, coolants should stay tightly sealed in an air conditioner's compressor, to be used over and over. If a carbon dioxide compressor does leak, it affects global warming much less than an equivalent amount of any of the super-greenhouse gases it replaces, and the quantity involved pales in significance compared to that produced by burning fossil fuels.

The Americans went back to the drawing board and the chemical laboratory, and began developing a different type of substitute molecule—still using fluorine but with much shorter atmospheric lifetimes and therefore almost no relative greenhouse effect (these gases are dubbed hydrofluoro-olefins in chemical language). History was repeating itself, with the Europeans favoring carbon dioxide and the Americans favoring synthetic chemicals, just as they had at the turn of the twentieth century. Some of this reflects the stronger influence of "green parties" in some European countries. It also seems to me to be cultural—with the Americans favoring technological solutions and new chemicals while the Europeans tend to be wary of new compounds made in the lab.

The political alignments for HFC reduction were the reverse of the Montreal Protocol's initial design. This time it was the Europeans who were pushing for faster environmental protection, not yet the United States. A spokesperson for General Motors urged a speedy change: "I encourage chemical companies to move as promptly as they can to allow us to evaluate new products that meet EU regulations," he said. "They need to comply with our development lead times and get involved now to meet the 2011 deadlines." California chimed in with an attempt at a state law to reduce HFC use in automobiles sold there as well.[6] Legal challenges by the automakers would delay California's implementation by many years, but the added threat of state regulation intensified the technology-steering pressure on chemical companies to speed up the development of alternatives.

In 2002, I took on a leadership role in the Intergovernmental Panel on Climate Change (IPCC), the organization that provides scientific assessments to the United Nations climate negotiators. I helped lead a special report on *Safeguarding the Ozone Layer and the Global Climate System* that was completed in 2005. One way to compare different greenhouse gases to each other is what scientists call carbon dioxide equivalent emission—it's how the warming effect of the emission of a particular gas translates into an equivalent amount of carbon dioxide release. Among the findings of that report was a graph showing how the CFC emissions compared to those of carbon dioxide over time. At their peak around 1990 just before the Montreal Protocol kicked in, the global CFC emissions translated into a surprisingly large value of carbon dioxide equivalents—about a third of that of carbon dioxide itself. This remarkable result brought home to me how prescient Ramanathan's warning was, and how good it was that the world was on track to eliminate CFCs, with emissions having already fallen by more than 75%. This was great for the ozone layer, and for climate too.[7]

A young Dutch scientist named Guus Velders was part of the international team that contributed to the report, and shortly after we finished, he and a group of colleagues got together for lunch during a scientific meeting and decided to take the next logical step—a step that was obvious to other scientists in 20-20 hindsight. Often the best and most influential scientific papers are obvious, and this was certainly an example of that. Velders and colleagues simply extrapolated the CFC emission forward from 1987 as if the Montreal Protocol had never happened, using the known rates of annual growth in the global market for these gases at the time. And a few percent growth every year compounded over a few decades generally packs a wallop, just as you may hope your retirement account will grow. By 2010 in a world without a Montreal Protocol, the CFCs would probably have reached ten gigatonnes per year (ten trillion kilograms) of CO_2-equivalent global emission. This number produced gasps of astonishment throughout the policy community, because it was five times larger than the two gigatonnes of greenhouse gas emission reduction by 2012 that was the hard-won goal of the Kyoto Protocol. The world suddenly realized that not only had the Montreal Protocol saved the ozone layer, it had done far more than any other treaty had yet achieved to stop climate change too.[8]

At first, the recognition of the Montreal Protocol's outsized role in climate protection was simply a cause for exuberant celebration by policymakers and scientists alike. But it quickly led to another conclusion that was more disturbing than joyful. About a billion and a half people live a high-income lifestyle today, and another six and a half billion live in poverty in the world's poorer nations. The environmental impacts of the rich world are immense, while those of poorer nations are far smaller. And we've already caused so much environmental damage and risk with such a small fraction of the world's population responsible for it—what happens when many more people rightfully increase their demands on the planet? While avoiding waste and conserving can save considerable planetary resources in the rich world, the benefits of technology for our health and well-being are also undeniable. The heart of sustainability is the challenge of how more equitable use of the world's resources and technologies can happen without creating multiple crises, simply because there are so many of us. If poorer nations develop using the same planetary resources (including fossil fuels, animal foods, minerals, and plants) as the high-income world has already done, then the Earth will not only get extremely hot, but many regions will become even more arid, and famine, water shortages, species extinction, and conflict can be expected to increase from their already-critical levels and reach catastrophic proportions. In short, to avoid dangerous levels of climate change we need not only to greatly reduce the release of greenhouse gases in the rich world, but also to find much better balances between our use of natural resources and global equity. Human development is sometimes framed as demanding better population control— but it is not so much the future growth in the world population that lies at the core of the issue in my view—because the large numbers of the impoverished who are already here have every right to development. Every convenience that the rich world currently enjoys requires energy and materials to make and use it, and generally results in greenhouse gas emissions: every toilet, every power line, every air conditioner, indeed, everything. People in the lower-income countries understandably seek to enjoy the modern miracles of refrigeration and air conditioning, as well as other conveniences. And their world is advancing quickly, with wealth growing by the day and along with it a more comfortable and safer lifestyle. Projections show that the wonders of refrigeration and air conditioning will become available to nearly all of their children and grandchildren by mid- to late-century. And if they acquire those benefits using the HFCs as we did using CFCs, then the HFCs would also play an outsized future role in future global warming, despite making only a small contribution today. By 2050, the HFC's annual contribution to global warming could even become similar to that of the CFCs before the Montreal Protocol, depending on how fast the lower-income countries develop, and with what kind of coolants. This was the second bombshell that Guus Velders dropped on the science and policy communities in a paper published in 2009.[9] Through the work of Velders and others, studies showed that keeping HFCs production going would threaten any chance of holding global warming below 2°C by 2050, which was by then the cherished goal of many an advocate for stopping global warming. On the other hand, replacing HFCs would shave up to 0.5°C off of the global warming that would otherwise occur by about 2050, making a significant contribution to climate change mitigation. While 2°C may not sound like much, we're already at about 1°C, and it's become clear that damaging climate changes are already happening, such as extreme heat waves that kill people. Even 0.5°C less warming will save many lives and reduce ecosystem damages.

While Velders's first paper on the CFCs was cause for celebration, the one on the HFCs caused an uproar. It raised the thorny issue of how to govern a type of substance introduced to save the ozone layer that now looked like it could seriously derail future efforts to safeguard the climate. Whose job was it to do something about this: the Montreal Protocol, or the Kyoto

Protocol, or some other treaty? Would even talking about this issue upset the ever-sensitive and delicately balanced climate negotiations? As the Copenhagen UN climate meeting of 2009 became tangled up in recriminations and failure to reach agreement, climate negotiation seemed not only to be ineffectual so far by comparison with the Montreal negotiations, but also to be hanging by a slender thread that might not hold up under controversy. As Mack McFarland of the DuPont company put it, "Everybody loved Velders's first paper, but lots of people hated the second one."

Among the first governments to put pressure on Montreal to take on the HFC issue was the Federated States of Micronesia, a nation in the equatorial Pacific that consists of more than 600 small islands. Like many island states, Micronesia's position on climate change stems from the knowledge that they are profoundly vulnerable to both sea level rise and changes in severe storms. While some of Micronesia's islands are high, many are small coral atolls that stand barely above the sea around them. Climate change therefore threatens their very existence.

Any nation that is a ratified party to the Montreal Protocol has the right to seek change, and in 2009, Micronesia submitted a "discussion paper" for the parties seeking to regulate the HFCs. By this time, the US chemical industry had caught up with the Europeans, and the United States had also jointly submitted a 2009 proposal with Mexico and Canada.

Other countries were not so eager. Many of the lower-income countries were fearful of switching coolants, deeply concerned about whether this would increase their costs. They also suspected that the wealthier nations might use the changeover as a chance to exploit them. For example, the rich might use the disruption to wriggle out of existing commitments to help the poor make the transition away from the CFCs. And by then many consumer refrigerators and air conditioning systems were coming out of China's booming manufacturing sector. Interference with their industry was unwelcome to the Chinese to say the least, and raised a plethora of trust issues regarding how any changes might affect their global markets. Further, in India as well as China, the domestic chemical industries had cherished hopes for an increased market share in the sales of substitutes for the CFCs, and switching targets was perceived as a threat to them as well. Negotiations continued at a glacial pace, with discussion papers repeatedly submitted for consideration year after year but making little traction.

A boost to progress arrived from an unexpected source. In 2011, the Food and Agriculture Organization of the United Nations (FAO) published an evaluation of the carbon footprint of food loss and waste. It showed that if food loss and waste were a country, it would be the world's third largest emitter of greenhouse gases, right behind China and the United States. Every step of food production, transportation, storage, and sales powered by fossil fuels leads to carbon emissions. Food production also leads to emissions of other important greenhouse gases, including methane and nitrous oxide. All of that is waste if the food spoils before it can be eaten. Meats, seafood, dairy products, vegetables, fruits, and even some processed foods will all last longer if kept cold during trucking, shipping, warehouse and market storage, and in the home.[10]

These "cold chains" not only limit food loss and waste, they are also essential for a number of medical applications. Many people first learned the term *cold chain* when vaccines for COVID-19 became available that require careful maintenance of temperatures below freezing. But cold chains are scarce to non-existent in much of the low-income parts of the world, particularly in rural areas. The FAO report revealed that overhauling the systems of food transportation and storage worldwide had an enormous potential to both mitigate climate change and provide more usable food to the world. Wasting less means feeding more people. And according to one source, food loss and waste leads to a staggering trillion dollars of economic loss per year. About half of the food loss and waste occurs in lower-income countries; the other half occurs in the middle- and upper-income world.[11] Phasing out the HFCs offered an opportunity to initiate a transformation, not only by using new coolants but also by making the systems that used them more efficient, and fostering adoption throughout the world. The HFC issue could therefore become part of an even broader and lofty goal of an improved food cold chain to reduce hunger while fighting global warming. Suddenly the Montreal Protocol had connections to a variety of other international efforts, not only within the United Nations (including the Zero Hunger Challenge and in the Sustainable Development Goals) but also in the activities of philanthropic organizations and international aid. In the words of Kevin Fay, executive director of the leading organization of chemical companies involved, the Alliance for Responsible Atmospheric Policy, "It's an opportunity that has a lot of good news associated with it."[12] Nevertheless, getting a global agreement on HFCs would not happen unless heroic efforts could gain the trust of skeptical lower-income countries.

While the US chemical companies had supported putting the HFCs into the Kyoto greenhouse gas trading basket, the climate negotiations became more and more tortuous under the Kyoto Protocol and its successor, the Paris Agreement signed in 2015, making the option of a switch to Montreal increasingly attractive. Progress toward the FAO's vision of greening and feeding the world through elimination of the HFCs coupled with

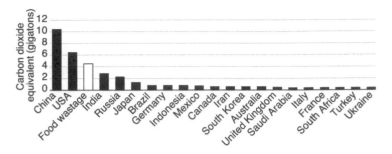

Total greenhouse gas emissions (in gigatonnes of carbon dioxide equivalents, expressed in mass of carbon) excluding land use, land use change, and forestry in 2011 for various countries, showing that if food loss and waste were a country, its emissions would be the third largest in the world.

modernized cold chain systems worldwide suited the industry's technological bent; it would also expand their markets. The chemical companies all knew that practical and affordable alternatives existed, including some industrial chemicals with insignificant greenhouse effects, such as hydrofluoro-olefins, or the carbon dioxide cooling that was already in use, especially in Europe. The governments would just need an international evaluation of the technical options, costs, and issues that both the high- and low-income countries could trust—and the ideal place to get one was Montreal. Some scholars have emphasized that the Montreal Protocol's technology assessments were a masterstroke: their innovation was to turn the chemical producers into partners rather than adversaries by drawing their experts into those evaluations. And according to Fay, that clever mechanism is what made industry realize that moving the HFCs into the focused and successful Montreal Protocol (and out of the sprawling and contentious Paris Agreement) was by far the best way for them.

American industry spent years working not only with the US State Department and the EPA to promote this structural change, but also with the governments of other countries and, more important, their counterparts in chemical companies in China and India, the countries that were, in Fay's words, "the blocking agents in getting this done." But the Montreal Protocol offered important precedents to help convince them. Its policies had carefully included a delay to allow the poorer countries more time for change than the richer countries. The price of the new products and technologies would decrease with economies of scale, as adoption spread first in the richer countries before the poorer countries would need to change. And the Montreal Protocol's Multilateral Fund had assisted the lower-income countries make the switch from CFCs by paying any incremental costs incurred by using substitute chemicals and products.[13] Those rules needed to be meticulously negotiated again if the HFCs were covered. There were also other concerns. The industry was well aware that future market growth of HFCs (or any alternatives to them) would mainly be in the poor countries of the world, even before Velders's paper highlighted how important it could be for mitigating global warming. China and India both had their own domestic chemical industries, busy making the HFCs as well as other compounds.

The lower-income countries had to be convinced that the move to phase the HFCs out in favor of something else "wasn't just a U.S. ploy to keep them out of the marketplace" as Fay puts it. Mack McFarland and Fay were foremost among the dogged travelers who racked up innumerable frequent-flier miles by making trips back and forth to these and other nations, endeavoring to build the needed global industry consensus. Ultimately, getting the lower-income countries on board through industry-to-industry discussions was the decisive factor in making policy change possible. Once the companies in those countries lost the fear of being left behind, they helped persuade their political representatives to put aside their fears of exploitation and go along too.

One might wonder what the chemical companies of the wealthier countries stood to gain. Industry had chosen the HFCs as the fastest route away from the ozone-damaging CFCs, but they knew all along that the days of the HFCs would be numbered because of their contributions to global warming. According to Fay, "Everybody knew the climate issue was not going away. This became the opportunity to come up with a path for a transition that was fair, that was sensitive, that was economically feasible." The industry also had witnessed how air conditioning had grown in the United States, enabling for example, the expansion of population and with it their domestic market in the Sun Belt. In 2010, US chemical companies had supported a phaseout of HFCs under the Waxman-Markey US climate energy bill, but after the dismal failures not only of that bill but also of later attempts to start the change at home under a climate umbrella, they decided national climate policy arena wasn't the right thing to work with. Instead the Alliance made the formal decision to pivot back to Montreal, with its demonstrated effectiveness and technology assessment panels.

In June of 2013, US President Barack Obama and Chinese President Xi Jinping met in California for their first summit.[14] Roadblocks that stymied his previous attempts to do something about climate change must have frustrated Obama, and with its strong industry support, the prospect of phasing out the HFCs was not an action he planned to pass up. As the two-day meeting drew to a close, the White House proudly announced that "for the first time, the United States and China will work together and with other countries to use the expertise and institutions of the Montreal Protocol to phase down the consumption and production of hydrofluorocarbons (HFCs), among other forms of multilateral cooperation."[15]

About a year later, Obama didn't achieve quite as much success in his first summit with Prime Minister Narendra Modi of India, but the two leaders warily announced a tiny bit of progress with a statement that "recognized the Montreal Protocol as the right forum for reducing HFCs"; they also established a joint task force for more discussions.[16] It was the first crack in official India's icy reception of the HFC phaseout.

After two more years and countless meetings by staff, the table had been set for a meeting of the parties to the Montreal Protocol in Kigali, Rwanda, in October of 2016 to amend officially the protocol and formalize an agreement for HFC reductions. India was among a small group of holdouts at the start of the meeting, but their opposition ended with an "exemption for parties with high ambient temperature conditions" that allowed them an added four-year delay. This was one of the most challenging parts of the negotiation, but was rationalized by the need to ensure that effective substitutes for the hottest nations would be available. The exemption also included Saudi Arabia, Qatar, and several other oil-producing nations that also dragged their feet. They were surely concerned about the intensive requirements for effective cooling in their oil industries and therefore wary of change. They also saw the obvious threat in the precedent created by any new and binding agreement to curtail climate change. But the pressure is intense at international negotiations, and often a special consideration such as this one can produce a breakthrough. India's negotiators also pushed hard for, and won, a measure requiring the richer countries to phase out even faster than the latter had proposed. The Kigali Amendment to the Montreal Protocol was historic, not least because it included

India's first move toward a specific agreement on reducing its own greenhouse gas emissions rather than claiming that all the responsibility had to rest with the rich countries.[17]

In addition to requiring that the HFCs be replaced in refrigeration and air conditioning, the agreement also requires analysis of overall emissions, including energy efficiency, sometimes referred to as a "life cycle analysis." Along with carbon dioxide and hydrofluoro-olefins, other substitutes such as hydrocarbons are also permitted in some applications. Despite being flammable, hydrocarbons can be good refrigerants in home systems and are widely used already in Europe. The TEAP would clearly have a great deal of work ahead of it for years to come, wrestling with the various options, their pros and cons, efficiency factors, and costs.

Secretary of State John Kerry traveled to Kigali as the standard-bearer for the United States. At a meeting a few months earlier, where he exhorted other nations to join the agreement, he proclaimed, "I want you to know there cannot be a shadow of doubt about where the United States of America stands on this. We are all in. We're committed to progress—progress here, success in Kigali, and to accelerating the effort to address climate change in our country and around the world."[18] He signed the agreement in Kigali with a flourish, even though he knew that any climate change agreement would be extremely tough to ratify in the US Senate. But he and Obama had plans to begin to address it via executive orders.

A few weeks later, in November of 2016, Donald Trump was elected as the forty-fifth president of the United States. Upon taking office, he promptly began to reverse a raft of Obama-era policies and actions. Progress on the HFCs was targeted for elimination, along with everything else related to climate change mitigation. US industry at first thought that Kigali still had a chance to be approved. But just when it seemed that they had managed to persuade Rex Tillerson, Trump's first Secretary of State, to support it, he was fired (for other reasons). Any near-term hope faded.

But Kerry had helped to craft a clever agreement that required ratification by only twenty countries—any twenty countries—to officially enter into force. The agreement entered into force in January of 2019, and it became increasingly clear that staying out would only hurt US industry in the long run, because the enforcement mechanism of the protocol is trade sanctions. The US industry dreaded the idea of being shut out of the export market.

Remarkably, in one of the most fractious eras in American history, bipartisan groups in both the US House of Representatives and Senate negotiated draft energy bills in October of 2020 that included domestic phaseouts of the HFCs, broadly consistent with the Kigali Amendment. As 2020 drew to a close, COVID economic relief measures, including the end of federal unemployment benefits and eviction moratoriums, were slated to expire. A government shutdown also loomed because extensions to avoid passing a 2021 government spending bill were sunsetting as well. A new bipartisan effort involving both the House and Senate carved out a compromise $900 billion COVID relief bill, combined with an omnibus government spending bill for 2021. In the words of one environmental activist, "the inclusion of an HFC phase-down in this [omnibus] bill represents the most significant climate legislation to pass in the US Congress in over a decade. It sends a signal that one of the largest contributors to climate change, the United States, is back at the global climate action table."[19] It also lit a bright beacon of change by demonstrating that bipartisanship on environmental issues was still possible in the United States, particularly when both industry and environmentalists strongly support proposed actions, as they did in this case.

After nearly a week of posturing, during which America's millions of COVID-unemployed endured a frighteningly uncertain Christmas, the Consolidated Appropriations Act 2021 containing both bills was signed by President Trump in the last days of 2020. The phaseout of HFCs had officially become part of US law, deftly aligning American domestic policy with Kigali's requirements.

When Joe Biden assumed the presidency in January of 2021, there was hope that an effort to ratify Kigali in America was coming, but there was also plenty of recognition that it would be hard. We had nothing to lose and everything to gain by ratification, since our own domestic policies already conformed to it, thanks to the omnibus bill. But there was still substantial resistance in the Senate—not because of the HFCs themselves, but because of a wide range of scarcely hidden agendas: ideological rejection of UN agreements of any kind as a threat to our sovereign rights to do everything we want (including dumb things), fear that regulation of HFCs would set a precedent for other greenhouse gases, objections to concessions given to India, and much more.

America's founders were skeptical of international linkages, and the Constitution makes transnational treaty ratification just about the hardest thing we ever do. The executive branch (almost always the State Department and the President) has the power to negotiate on our behalf as a nation, but the Senate must approve a treaty by a whopping two-thirds majority or better (the largest majority of any Congressional action) and only then can the president sign to officially conclude the ratification process. How on Earth is that going to happen anytime soon, regardless of that great 2020 energy bill? I felt sadly doubtful that it could happen in less than a decade amid today's culture wars. Even when industry began a concerted push to persuade reluctant senators to step up, I remained skeptical of success, a very unusual twist compared to my generally optimistic views.

Just when you might have thought such agreement was impossible, a miracle occurred. In September of 2022, the US Senate ratified the Kigali Amendment with votes to spare, 69–27. Republican Senator John Neely Kennedy was the lead sponsor. Kennedy hails from Louisiana, the state that houses the Honeywell and Mexichem Fluor chemical companies, which produce alternatives to the HFCs, and certainly stood to gain from ratification. As his co-sponsor quipped, "it's not every day that you have a full court press from the business community and are joined with a full court press from the environmental community."[20] You may not know that the United States has a stake in 75% of the world's air conditioning equipment, or that there is an Air Conditioning, Heating, and Refrigeration Institute in Arlington, Virginia, another of the many lobbying organizations that encircle Washington. When Kigali was ratified, its president trumpeted that "the Senate is signaling that Kigali counts for the jobs it will create; for global competitive advantage it creates; the additional exports that will result, and it counts for U.S. technology preeminence," adding that demand was "exploding." Kigali clearly demonstrated that climate change itself is not always a hot-button issue, even in the United States of the twenty-first century.

Kigali shows that industry can be extremely important in environmental success. That mattered a lot here, along with leadership by officials at the helms of major countries (including ours) who set up the agreement in the first place. Kigali also had the precious gifts of the precedents, institutional structure, and level of international trust already set up in Montreal. The fact that climate change is becoming more and more personal and perceptible probably helped, and the switch to other coolants is a pretty easy thing to do.

Kigali is basically a triumph of the practical. Options are available. The connection to food waste and reducing hunger around the world was also a big plus that many people could find motivating, boosting public support. And it built on the strong citizen's history that came before it in the Montreal Protocol, starting with consumer boycotts of spray cans containing CFCs that kickstarted Montreal to success in the first place.

The chemical companies are not angels. They've done some horrendous things in the past and still do. You can find plenty of sites both in the United States and abroad where toxic chemicals have poisoned entire communities. While slow to step up at first on accountability for their products, they learned the lesson over the last half century that sometimes it's better to work with city hall than to fight it.

Surely a leading factor in Kigali's success is that the halocarbon chemicals industry is very far from being an economic titan—the global market was estimated at a few tens of billions of dollars in 2020, while the fossil fuel industry is roughly a thousand times bigger, in the tens of trillions. For a long time thinking about this made me feel helpless. But then I began to reflect on precedents. Has a big mega-industry ever faced up to an environmental challenge?

3

Smog

Residents of Los Angeles had long been aware of the smog that
dimmed their sunlight and fouled their air, but for Marge Levee,
the perceptible became personal in 1958, when a severe asthma
attack sent her two-year-old daughter to the emergency room.

This was the era of stay-at-home mothers and "Father Knows
Best," but Marge had had it with suffering in silence. She gathered a group of equally enraged, equally well-to-do LA moms
to form Stamp Out Smog (SOS), a group that met regularly in
the Levee's Beverly Hills living room. And these women weren't
just angry and well-heeled; they were angry and well-connected.
Marge was the wife of a film producer and talent agent. Others
were married to prominent entertainers. These included the
wives of Art Linkletter and Robert Cummings—huge television
stars at the time.

Steeped in the ways of Hollywood, the members of SOS knew how to maximize attention, and it was not by writing letters to editors or gathering signatures on petitions. They took to the streets wearing their pearls and white gloves, often bringing along their little kids sporting gas masks. At a time when "social media" was limited to rotary phones and chain letters, they allied with more than four hundred other groups, building an extensive organization of garden clubs, unions, health organizations, even chambers of commerce. In an era of soul-crushing sexism and social conformity, this strident activism from wives and mothers expressing concern about the welfare of their families managed to come across not only as socially acceptable, but as urgent.[1]

According to a Palm Springs newspaper, "SOS probably packs the most concentrated and potent feminine determination ever directed towards achievement of a common goal in the history of our Golden State. Woe betide the hapless politician or industrialist who fails to show complete cooperation."[2] But the pieces necessary to solve the smog puzzle didn't magically fall into place. SOS was a brilliant step, but a genuine solution would take a much broader effort, because the problem had a much broader scope and much deeper roots than bad air on Wilshire Boulevard. Most Americans probably know the lines inscribed on the Statue of Liberty in New York harbor: "Give me your tired, your poor, your huddled masses yearning to breathe free." What they probably don't know is that in 1886, at its official inauguration, the statue of Lady Liberty was all but obscured by heavy pollution created by a regatta of coal-powered steamers.[3]

Tracing the history of human-induced air pollution back to its origins would be a very long journey. Suffice it to say that the fires Neanderthals used to heat their caves left behind greasy soot that coated the walls, and showed up in mummified lung samples analyzed by today's archaeologists.[4] Then again, what happened in the cave stayed in the cave. Outside, lay a pristine atmosphere that would remain that way for tens of thousands of years. It was when cities developed that smoke from kitchens began to drift through streets and choke all and sundry, from antiquity to the modern era.

Medieval London was known for its exceptionally dirty air well before 1377, when the city regulated the minimum heights of its chimneys in an attempt to address its growing misery with the approach of "dilution as the solution." Surely, the vast Earth could handle whatever refuse humans chose to throw at it—as long as city dwellers could find ways to spread it out. This ill-advised strategy would be practiced many times through the centuries, despite evidence that if it helped at all, the improvement was both minimal and transitory.[5]

As London's population grew, increasing levels of pollution quickly overwhelmed the benefits of higher chimneys. As the air became dirtier still, the chimneys had to be lifted again, over and over and over. Ultimately, the lofting of pollutants from very tall chimneys (and later industrial stacks) in London and many places worldwide served mostly to deposit damaging acid rain on faraway forests and lakes.

With the coming of industrialization, coal-powered steam engines belched soot, nitrogen oxides, carbon monoxide, and other pollutants in staggering amounts. Textile factories and metalworking foundries transformed many previously bucolic areas into grimy, gritty slums. By the time of Charles Dickens, London was infamous for its toxic mix of soot and sulfur dioxide from coal burning, which combined to form traditional winter "smokes." Sometimes the pall grew thick enough and lasted long enough to kill people (about 650 in 1873–74; another 800 in 1892). The unhealthy brew of smoke and fog inspired a new English word: *Smog*.[6]

Little more than a half century later, Londoners coughed and wheezed their way through the "Big Smoke" of December 1952. Initial estimates put the number of fatalities between 3500 to 4000, but as the deaths continued into spring, the toll was raised to include roughly 12,000 victims.[7] London was one of the most polluted cities in Europe because of its heavy use of high-sulfur coal, but it was far from the only locality suffering from foul air. At least sixty people had died in the Meuse Valley of Belgium during a smog event in 1930. But in this era, regulators and the public considered air pollution a local matter and were mostly concerned with the visible black smoke as a "nuisance" pollution.[8]

In the United States, Chicago passed the first American smoke ordinance in 1881, aimed only at reducing the thick black clouds billowing out from its growing industrial base. There was no scientific understanding yet to show the harm from invisible gases, and that these could spread far and wide from those belching smokestacks of which industrial cities were so proud. Moreover, enforcement of regulations was generally weak enough that industry hardly noticed.[9]

Soon to become the capital city of a new type of US-style smog, Los Angeles experienced its first major episode in 1943, with a pall of smoke obscuring the Sun and residents suffering through coughing fits and asthma attacks. But the event that made the risks of lethal air quality perceptible to most Americans took place five years later, in Donora, Pennsylvania, a steel town along the Monongahela River. Nobody seemed to be affected in Pittsburgh, just thirty miles northwest, but in Donora, where American Steel and Wire had built a zinc works, about 40% of the town was made sick, seventeen people died, and thousands were hospitalized. Neighboring towns experienced fog but no deaths or illness, so citizens suspected the zinc factory and clamored for a local smoke-control ordinance. The zinc works fought back, employing deflection tactics that would become standard in the industrial playbook: deny culpability, then call for conclusive evidence to escape admitting that a problem exists.[10]

Donora, like Los Angeles and Belgium's Meuse Valley, lies in the bottom of a topographical bowl that allows air to become trapped during stagnant weather conditions. American Steel and Wire turned to the Kettering Laboratory at the University of Cincinnati, which had been largely funded by the General Motors corporation at its inception, for an investigation. The Kettering Laboratory was led by Dr. Robert Kehoe, who was already known for his defense of the use of leaded gasoline and had a sweeping view that metal pollution was generally a natural part of life, making him industry's obvious choice. But civic pressure mounted against the industry inquiry, forcing the US Public Health Survey to start its own investigation. Measurements of air pollution and metals began in earnest around the area for the first time.

In early December 1948 the Survey's personnel conducted a door-to-door survey, but the interviewers were frustrated by evasive responses. They noted that many of the residents worked at the plant and appeared fearful of incurring its wrath. Ultimately, both the Public Health Survey and Kettering concluded that the Zinc Works could not be held responsible. They cited geography and weather as the major culprits, despite the fact that such weather itself, while perhaps unpleasant, doesn't kill or sicken so many people.[11] The Donora incident stimulated the first serious discussions of US legislative proposals on air pollution, as well as sponsorship of independent research at the state and national levels. Perhaps most important, the news from Donora informed Americans throughout the country that air pollution had to be considered a health risk, well beyond the big cities.

By 1950, burning coal was producing enough "nuisance" smoke that eighty municipalities, many of them in cold climates such as Ohio, Pennsylvania, New York, New Jersey, and Massachusetts, had adopted some form of local pollution laws. Still, in outlying areas, nothing was being done. Only two American counties (and not a single state) had comprehensive air pollution control measures.[12]

For the most part, city regulations focused on the then-novel requirement that plans and specifications for fuel-burning equipment be approved before installation and that construction and operating permits be required for sources of dense smoke. While such measures helped to reduce the belching black fog, dust, and fly ash from particular sources, they did little to limit the increasing levels of pollutants, or their impact outside urban centers. Many of the measures taken were logical enough, but some were absurd, such as the requirement in one New Jersey county that the rotten-egg odor of sulfur gases coming from a chemical plant be dealt with via a deodorizing chemical released to mask the smell.[13]

Before the 1940s, when Southern California had its first smog crisis, Los Angeles was seen as a kind of Shangri-La, its sunshine, pleasant climate, and clean air luring folks from points east with the promise of a new beginning, low-cost housing, and an escape from ice and snow. Between 1920 and 1940, the appeal of Los Angeles generated faster growth than anywhere else in the nation, with the population of LA County nearly tripling from 1 to 2.7 million. When the United States entered World War II, LA's industrial base rapidly increased as well, with burgeoning manufacturing, especially in defense industries.[14] But the economic boom was only one factor in the area's air pollution.

LA is surrounded on three sides by mountains, with the ocean as the border to the west. A sea breeze routinely brings cool air from the Pacific into the bowl-shaped basin, where it sits near the ground, capped by warmer air at higher altitudes. Since hot air rises and cold air falls, this altitude profile makes it very hard for polluted air near the ground to get out of the bowl. Unless there is a stiff wind, the air is unable to cross the mountains, which allows pollution to build up, particularly in summer and fall. Winds do switch direction to flow away from the land to the ocean at night, but they are weaker on average, so the pollution lingers and compounds, day after day.

As early as 1542, explorer Juan Rodriguez Cabrillo had noticed that the smoke from campfires rose no more than a few hundred feet before spreading out in a layer of haze, obscuring the view of the mountains even then. He dubbed San Pedro Bay "La Bahia de Los Fumos," the Bay of Smokes. By the 1940s, La Cuenca de Los Angeles, "the basin of angels," was anything but angelic, with the worst air quality in the nation. Smog events routinely blocked the previously majestic views of Catalina Island and the mountains, even causing "daylight dimouts" that forced movie crews to stop outdoor filming.[15] Smog certainly had become quite perceptible.

The millions who had flocked to California for its natural beauty and healthful environment were appalled. Harry Chandler, the influential publisher of the *Los Angeles Times*, pushed his paper to report extensively on the problem, and he took the unusual step of establishing a Citizens Smog Advisory Committee. While the chambers of commerce in nearly every eastern and midwestern city vigorously opposed research to find the causes of pollution, and even more stridently howled about regulation, LA's chamber was entirely different.

Even for conservative business leaders, this highly perceptible problem was also personal. They knew all too well that the area's prosperity (particularly its lucrative real estate industry) was tied to its beauty and the promise of escape from the soot and smoke of cities farther east. An unusual coalition of the public, media, and business leaders began to press for answers, ultimately forming a united front that trained its attention on getting action from policymakers.[16]

During LA's benchmark Black Wednesday smog event in 1943, greasy films formed on furniture, the paint jobs on cars blistered, and the population struggled through an extreme daylight dimout. City officials suspected a chemical plant that made butadiene for rubber production, an important part of the war effort. Following a political battle, devices were installed to control the plant's emissions, but any relief for Angelenos was short-lived. The plant's odors ceased when the control measures were taken, but smog events did not.[17]

In 1945, responding to public and media pressure, LA created a Smoke and Fumes Commission. They attempted the standard approach of further restricting coal burning, but that didn't work the way it had in the cities of the Northeast or in London. LA's smog remained stubbornly thick, and visibility continued to decline. Emergency rooms were overwhelmed by bronchitis attacks, and the incidence of childhood asthma grew. Later studies would document an association of increased emphysema and deadly heart attacks with the worst smog episodes.

Advocacy had led to the creation of committees and commissions, but to find truly effective solutions for LA's foul air, the city's elders, in another pattern that will become familiar, called on science to develop a better understanding of the complex, causative factors. Smog did not appear to stem mainly from burning coal, or from the butadiene plant, so what was the culprit?

The *Los Angeles Times* brought in an air pollution expert from Washington University in St. Louis to do a study. He concluded that the growth of industries around LA was responsible for the county's air-quality problems, and he focused on chemical companies, railroads, refineries, and companies that made soap, paints, and foods. He explicitly dismissed much of a role for LA's explosive growth in the number of automobiles; instead, he recommended strictly enforced controls on smokestacks, factories, backyard incinerators, and other stationary forms of pollution.

While these measures ultimately achieved little, an important upshot of the study was the realization that the problem had to be addressed throughout the LA Basin, which meant including more than forty outlying towns. The city of Los Angeles asked the state of California to set up a Los Angeles County Air Pollution Control District to plan and carry out the regulations and empower the police to enforce the law.[18] But the question remained as to what, exactly, should be regulated. California smog appeared strangely different from that encountered elsewhere, which meant that the investigation needed to turn once again to science, to get down to the level of basic chemistry.

Arie Haagen-Smit, a professor at the California Institute of Technology in Pasadena, was an expert on the structure and properties of the molecules that give rose oil and other perfumes their pleasant smells, and how these molecules form in flowers, fruits, and other botanical sources. "Haggy" (to his friends) was born in Holland in 1900 and completed his doctoral thesis at the University of Utrecht on *sesquiterpenes*, one of the compounds that account for the distinctive aroma of citrus fruits.[19] In 1949, Haagen-Smit had analyzed the chemicals that give pineapple its taste and fragrance. He concentrated the aromas of pineapple by passing fragrant air through a cooled tube to make its gases condense into "freeze-out traps" to isolate the compounds. Arnold Beckman, a former Cal Tech professor who had become a leading maker of scientific instruments, suggested that Haggy apply his technique to examining the smog outside his laboratory window.[20]

Haagen-Smit proceeded to pass five hundred cubic meters of polluted air through his freeze-out traps, which yielded a few drops of "brown, gooey, smelly stuff." He then initiated the chain of breakthroughs that ultimately revealed the source of LA's woes. Haggy's analysis showed that the gooey stuff contained carbon, hydrogen, and some oxygen. This gooey mixture pointed to hydrocarbons and not sulfur as a key culprit in smog. Compounds containing carbon are called "organics" because the carbon atom is a pervasive building block of life, but many hydrocarbons don't come from natural sources. This appeared to be the case with the hydrocarbons found in Haggy's brown goo, which prompted speculation as to where they came from. Haggy and Beckman suspected that an important source was leakage from refineries, which understandably aroused the fury of the local oil industry.[21]

Haagen-Smit could have published this study and gone back to his uncontroversial and very productive research on flavors and smells, but Beckman engineered an interaction that he knew would inspire the "stubborn Dutchman," as he referred to him, to double down.

Western Oil & Gas Association had already funded an effort at the Stanford Research Institute, and a chemist there was familiar with the organic compounds in Haagen-Smit's goo. This chemist argued that these "peroxy-organic compounds" couldn't be the cause of smog because they didn't have a smell, nor did they irritate the eyes. So, the wily Beckman invited that chemist to come down and give a lecture at the Cal Tech chemistry department. In his talk, the expert went so far as to say, "It's unfortunate that a good chemist like Haagen-Smit could be misled." Beckman was sitting next to Haagen-Smit at that moment, and said he "could almost feel Haggy's blood pressure rise." Haggy muttered "I'll show them," and left the room.[22]

It's very likely that Haagen-Smit already suspected what might explain the peroxy-organics, as well as the important role of a very smelly chemical that might have caused them to form in the first place. That special compound was ozone, which gets its name from the Greek word "ozein" (to smell). Its odor is pungent and acrid, somewhat like bleach. You may have smelled it even in clean air after an intense lightning strike, because lightning can break the strong chemical bond of the oxygen in air, allowing it to form O_3, a.k.a. ozone. And ozone is a very reactive gas. But even if, as seems plausible to a chemist, it was ozone's interaction with hydrocarbons that was forming the peroxy-organic compounds in LA's air, that still didn't solve the mystery of where the ozone was coming from in the first place.

Since the mid-1800s, scientists had known that pollution damaged plants and crops, and by the 1940s, the forage crops, grasses, flowers, ornamentals, and leafy vegetables of the Los Angeles valley were suffering mightily under the growing load of smog. The symptoms of LA's damaged vegetation, however, were described as a metallic silvering or bronzed spotting. This spoiling was different from that found in the Northeastern American cities and other areas, where the local smog could be traced to coal. The worse the air pollution became, and the more the visibility decreased, the more damage was done to crops, including California staples such as lettuce, endive, spinach, and oats.[23]

Haagen-Smit formed a team of plant researchers drawn from Cal Tech and the University of California at Riverside to study the damage and track down its cause. The appearance of the stricken leaves was not reproduced in experiments exposing plants to sulfur dioxide, ammonia, or other chemicals suspected in plant damage elsewhere. So, the scientists grew the spinach, sugar beets, endive, oats, and alfalfa plants in a chamber filled with smog-free air that had been filtered through activated charcoal. In another chamber, they grew the same plants in the unfiltered stuff that passed for air in Pasadena. In the smog chamber, the characteristic silvering occurred within days, and stunted plant growth to boot, while the plants in the other chamber stayed robust and healthy.

The team then analyzed the individual chemical elements—pentane, pyridine, acrolein, and many more—within the smog, and fumigated more plants with each in turn, looking for the combination that would cause all of the different crops to show their common "smog" response. Only when the plants were exposed to a mixture of hydrocarbons (including gasoline) and ozone (produced artificially) did all the crops shrivel and silver as they did along Southern California roadways. Which still left a mystery for the scientists to solve. There were plenty of potential sources for hydrocarbons in Los Angeles's automobile emissions, refineries, and other industries, but the hydrocarbons alone didn't cause the telltale damage.[24] So, where did the high levels of ozone critical to crop damage come from?

Rubber auto tires offered a clue, as well as a new exploratory technique. Tires had long been known to crack and age faster in the LA area than elsewhere, but no one knew why until 1950, when Haagen-Smit cleverly exploited this feature in a new study. In the laboratory, he and his colleagues stressed standardized pieces of thick rubber by folding them, then measured the amount of time required for the rubber to crack under exposure to different amounts of ozone. They also exposed the rubber to other chemicals found in smog, but it was the ozone that proved to be rubber's nemesis. The more ozone, the faster the rubber deteriorated. Not only had they explained the tire damage, they had found a new, simple, and easily-portable method for figuring out how much ozone was in the air.[25]

Haagen-Smit took his new method outside to measure the dramatic buildup of ozone during a smog episode, and it took only a few minutes for the rubber to crack. Haggy speculated that the key extra spice needed to make that ozone might be nitrogen oxides—another pollutant produced by automobiles and industry as a by-product of burning fossil fuels at high temperatures, combined with the action of Pasadena's bright sunlight.[26] While ozone in the upper atmosphere protects us from ultraviolet light, it's very bad news at ground level—high levels of ozone in LA's air were a key part of its smog problem, wreaking havoc on plants and humans alike.

LA's love of the automobile, combined with its abundant sunshine and its location in a basin ringed by mountains, made it the perfect setting for a new kind of twentieth-century pollution, different from that seen anywhere else. But there were skeptics, chief among them Harold Johnston, a highly respected physical chemist who had extensively studied the speeds of many different chemical reactions involving nitrogen. At a seminar at Stanford in 1951, Johnston discussed Haagen-Smit's remarkable new findings about California smog. He knew the speeds of those nitrogen reactions well enough to say with confidence that the action of the sun on nitrogen oxides—on its own—was too slow to explain smog events.

A representative of the oil and gas industry was sitting in the back of the room, and he approached Johnston after the talk, seeking a review of Haagen-Smit's findings. He arranged for Johnston to give an informal talk to the board of directors of the Stanford Research Institute, where a great deal of work funded by the oil and gas industry was taking place. Johnston recounted that "the oil people were vociferous that Haagen-Smit was all wrong, that he was only a publicity-seeker. Automobiles and gasoline couldn't have anything to do with it. They really opposed him on a personal level." Johnston added, "I didn't say he was wrong all the way. I just said he was wrong in one particular reaction." But the industry men persisted, and Johnston agreed to do the study.[27]

Haagen-Smit had already gone beyond the idea of the nitrogen chemistry acting alone to make the ozone, working out a more complicated mechanism. His idea combined nitrogen with hydrocarbons in a chain of reactions involving chemicals formed as the hydrocarbons decomposed in sunlit air. Those breakdown products (peroxy radicals) could then react with nitrogen and sunlight to make ozone, and that in turn could react with hydrocarbons to make the goo he had found in his traps. In a paper published in 1953, he argued that the formation of ozone occurred through this alchemy of auto exhaust and sunlit air trapped in the LA Basin.[28]

Industry's hopes that Johnston would come to their rescue went up in smoke. When Johnston went through Haagen-Smit's papers documenting the chemistry in more detail, he "rapidly concluded that Haagen-Smit was a genius!" When he reported that he basically agreed with Haagen-Smit's findings, that was the end of any Johnston contracts or consultancies with the oil barons.[29]

Johnston went on to publish a series of important papers of his own, examining exactly how those peroxy compounds were formed and how fast their reactions were, and his work would become another of the cornerstones of scientific understanding of smog chemistry. "The detailed mechanism took years to work out," as he put it, but he always recognized Haagen-Smit as the leader and pioneer, and the two communicated with one another about their work.[30] By the mid-1950s, Haagen-Smit's theory combining sunlight and chemistry in "photochemical generation of smog" had been widely confirmed.

Ari Haagen-Smit making smog perceptible to his audience during a lecture.

As public and media concern grew—much of it generated by Marge Levee and her Stamp Out Smog group—the automakers slowly came around. In 1958, Haagen-Smit took leave from Cal Tech to become the director of research at the Los Angeles County Air Pollution Control District. One day, a colleague found Haggy chuckling and asked what was so funny. He replied, "Today I had three vice presidents from Ford Motor Company in my office. Last year, I would have had to go to Detroit to see them, if I could have seen them at all."[31]

In 1960, the California state legislature passed the Motor Vehicle Pollution Control Act, the first US action to set specific standards for controlling pollution from automobiles. The Motor Vehicle Pollution Control Board was its governing body, and Levee was its sole female member. Because, in the words of one official, "These SOS women are afraid of no one," the public pressure in California would continue to be a force to reckon with in pollution control throughout the decade.[32] First the Control Board certified new pollution-control devices to be put on all new and used vehicles. And in 1964, California set limits on allowable tailpipe emissions for all new cars to be sold in the state, starting with the 1966 models.[33] By this time, Californians were not alone in their air-quality concerns.

While Los Angeles was, and remains, the national poster child for America's air pollution problem, expanding industry and car ownership led to increasingly dirty air in much of urban America. No longer were the eastern cities' problems controllable by monitoring smokestacks and lifting up chimneys. In 1953, about two hundred people died[34] in New York City's first, full-scale smog event.[35] Serious air pollution was now widely perceived, but it would take another decade for the problem to become personal enough, nationwide, to prompt concerted action. The 1950s were focused on recovering from the traumas of the Great Depression and World War II, avoiding a nuclear showdown with the USSR, and making the most of America's new-found prosperity.

In 1960, the election of the young and charismatic President John F. Kennedy heralded a new era of optimistic engagement, which he called "The New Frontier," much of it focused on new technology, with the signature goal of putting a man on the moon. Kennedy's assassination three years later crushed the national spirit and fueled endless speculation about what else he might have accomplished—including how he might have dealt with the Civil Rights Movement, pollution, and the other challenges of the sixties. But when JFK's vice president, Lyndon Baines Johnson, replaced him, LBJ lost little time in making plain his intentions with regard to social justice and the environment.

Chief among his allies was the junior senator from Maine, Edmund Muskie. If you are an American who enjoys reasonably clean air and water, you owe a debt to Muskie, who became a US senator from Maine in 1959, after two terms as its governor. He served in the Senate for more than two decades and had an enduring interest in environment. He would become the primary architect of both the US Clean Air and Clean Water Acts in the 1970s. While there were numerous policy entrepreneurs in American environmental history, Muskie is its leading environmental policy kingpin.[36]

Assigned to chair a new, special subcommittee on air and water pollution, Muskie would skillfully use the platform throughout the decade. His first effort was to hold hearings to attract national attention, then deploy congressional staff and resources for intensive studies to lay the technical groundwork for the 1963 Clean Air Act, passed one month after Kennedy's death. Muskie then shifted his efforts beyond the Beltway and into the country at large, conducting air pollution hearings in Los Angeles, Denver, Chicago, Boston, New York, and Tampa that addressed local concerns, and attracted even more press to galvanize the public.[37]

In May of 1964, LBJ struck a blow for clean air. In his commencement speech at the University of Michigan, he announced his plan for the comprehensive social programs known as the "Great Society." Referencing our pride in being "America the beautiful," he noted that such beauty was in danger, saying that, "The water we drink, the food we eat, the very air that we breathe, are threatened with pollution."[38]

Johnson's speech resonated with intensifying calls for change and social movements flowing across the country at the time. Americans had absorbed Martin Luther King Jr.'s "I Have a Dream Speech" during the March on Washington, the Birmingham church bombing, and the killing of civil rights leader Medgar Evers. On the cultural front, we had Beatlemania and the further

"British Invasion," along with the beginnings of what would become the era of "sex, drugs, and rock 'n' roll." On a deeper level, we had the publication of *The Feminine Mystique*, Betty Friedan's best-selling manifesto that captured the yearning of women everywhere for new and more expansive roles in public life.

It was appropriate, then, that one of Muskie's most effective environmental allies was the president's wife, Lady Bird Johnson, the shrewd tactician behind the Johnsons' fortune in radio and television. As the focus for her work as First Lady, she chose the seemingly innocuous theme of "beautification."[39] But her "ladylike" advocacy of flowerbeds and roadside cleanup appears to have been the thin edge of a clever wedge to open up a much larger discussion of environmental problems, including air pollution. She subtly linked the two with the statement, "Beautification means our total concern for the physical and human quality we pass on to our children and the future."[40]

Lady Bird became a role model who further energized women's organizations that, like the SOS group in Los Angeles, had been leading the environmental effort at the local level. In an era when women were bristling against the expectation that they should focus exclusively on home and hearth, these activists expanded the concept of home to include the "home planet."

Ultimately, though, hostility to the Vietnam War became the lever that pried loose conventional ways of thinking. As young people took to the streets to express their frustration with the war, their rage extended to the corporations who greased the wheels of the war machine and to the ways in which industry insisted on dominating the natural environment rather than living within it in a healthy way.[41]

The destabilizing energy of these social movements came to a head in the fateful year of 1968, when tactical failure in Vietnam, along with a record number of US casualties, led President Johnson to announce that he would not seek re-election. This was followed only a few days later by the assassination of Martin Luther King Jr. and then the assassination of Robert F. Kennedy in June 1968, after which anger boiled over into violence, with rioting and looting all summer, capped by the chaotic and disastrous Democratic National Convention in Chicago. The net result was that the fearful and exhausted electorate retreated to supposed "law and order" by electing Richard M. Nixon.

Eight days after his inauguration on January 20, 1969, an offshore oil rig began to blow apart off the coast of Santa Barbara. In little more than a week, it would spew about 100,000 gallons of crude, setting a new record for oil pollution incidents and killing thousands of sea birds and marine mammals. Television brought the death throes of oil-soaked wildlife into homes, not just in California, but all across America, pushing the environmental movement to even higher levels of outrage and prompting more protest marches.[42]

Reacting to the public mood, Nixon made promises to improve environmental protection: "What is involved is something much bigger than Santa Barbara . . . [it] is the use of our resources of the sea and the land in a more effective way, and with more concern for preserving the beauty and the natural resources that are so important to any kind of society that we want for the future. . . . I don't think we have paid enough attention to this. . . . We are going to do a better job than we have done in the past."[43]

Six months later, the heavily polluted Cuyahoga River, long known for "oozing" rather than flowing from Akron to Cleveland, Ohio, burst into flames.[44] Meanwhile, US forces in Vietnam, already setting the jungle ablaze with napalm, now began spraying a new and devastating chemical from the corporate giant Monsanto: Agent Orange. This herbicide, meant to destroy the jungle canopy where Viet Cong insurgents might hide, destroyed farmland indiscriminately, and led to horrific birth defects among Vietnam civilians. It was also later associated with cancers, Parkinson's, and other diseases in US soldiers.[45]

With persistent images of seemingly pointless bloodshed in a tortured and denuded landscape, the war in Vietnam came to be viewed not as a war against the advance of communism, but as a war against Nature, and against common sense. The peace movement and the environmental movements coalesced, particularly among student activists. John Lennon's anthem, "Give Peace a Chance," morphed into "Give Earth a Chance," and the stage was set for the most sweeping environmental legislation in American history. Smog was no longer smoldering on the back burner of public attention. Choking children, deaths, and widespread anger, activism, and unease had created a hot crisis.[46]

Senator Gaylord Nelson of Wisconsin captured the moment with the inspired idea of a nationwide teach-in, focused on the ways humanity was destroying the planet. Interest in the event quickly mushroomed, overwhelming Nelson's small Senate staff, and he wisely hired a twenty-five-year-old activist at the Harvard Kennedy School named Denis Hayes to coordinate the effort. As Hayes organized what became known as Earth Day, the movement escalated into a nationwide protest that would eclipse anything that had come before.[47]

In 1969, the *New York Times* reported that the environment was overtaking Vietnam as the number one concern in colleges and universities throughout the nation. Nixon felt pressure to respond. He also feared that the Democrats' historic strength on environmental issues would threaten his re-election. As a result, his State of the Union address on January 22, 1970, sounded more like a clarion call from the Sierra Club than the sentiments of an unprincipled president who would be driven from office three and a half years later by the Watergate scandal. From the podium, Nixon proclaimed, "The great question of the seventies is, Shall we surrender to our surroundings, or shall we make our peace with nature and begin to make reparations for the damage we have done to our air, to our land, and to our water? . . . Restoring nature to its natural state is a cause beyond party and beyond factions. . . . Clean air, clean water, open space—these should once again be the birthright of every American."[48]

Two days later, Ed Muskie, who had been the Democrat's vice-presidential candidate in 1968 and was clearly the most likely challenger to Nixon in 1972, held his own press conference. He acknowledged Nixon's rhetoric but argued that the specifics of Nixon's suggested changes would be woefully inadequate.[49] Muskie began a yearlong frenzy of statements, speeches, hearings, articles, and appearances to push for a true revolution in environmental policy, which came to fruition in the Clean Air Act Amendment of 1970, a watershed moment in American environmental history.

On February 1, 1970, Muskie was a guest on *Meet the Press*;[50] the next day, he was on the radio; then he started a hearing on power development and the environment and introduced the Omnibus Water Quality Bill to the Senate. On February 3, he issued a statement critiquing the administration's environmental budget request.[51] Muskie's appetite for advancing environmental protection seemed limitless.

Near the end of February, Muskie published an article in the *Ladies' Home Journal* entitled "Our Polluted America: What Women Can Do." He wrote, "We are disgusted by the dirty air, rotten odors and vile water. . . . Many housewives and professional women already belong to the Citizens Crusade for Clean Water, an ad hoc coalition of dozens of organizations, including 157,000 members of the League of Women Voters, whose 1,275 local leagues have called upon women to testify, write letters,

Women and children protesting against smog outside a hearing room at the Los Angeles County Board of Supervisors, March 16, 1961.

distribute fact sheets, attend public meetings and spark community dialogue about pollution. . . . I would like to suggest a plan for all women who are not yet involved [to join] in programs to stop environmental pollution. . . . I believe she will soon find herself head over heels in a new, life-long romance—with all of the elements of life on earth. And that, it seems to me, is what the power of a woman should be all about."[52]

While Muskie exhorted women to do more, Gaylord Nelson and his team exhorted the young. The Moratorium to End the War in Vietnam in October 1969, had attracted roughly two million people. The first Earth Day in April of 1970 drew an astounding twenty million nationwide.[53] In cities and towns across the country, the young reached past their campuses to children's schools, churches, and beyond to demonstrate, wear flowers, and sing and march in front of the offices of polluting industries and along the shores of polluted rivers. That such a turnout was possible in 1970, in an era of snail mail and pay phones, attests to the activism and devotion of Hayes and his networks.[54]

A highly experienced scholar of public opinion subsequently described with astonishment "the unprecedented speed and urgency with which ecological issues have burst into American consciousness." For several years, pollsters at Opinion Research Corporation had systematically asked Americans whether the problems of air and water pollution in their area were "very," "somewhat," or "not very serious." Results for 1966 stood at 48% answering "very" or "somewhat." This crept up to 53% in 1967, then 55% in 1968. In June of 1970, immediately after Earth Day, "very" or "somewhat serious" catapulted to 69%.[55]

Many years later, Denis Hayes related a story he'd been told by White House Counsel and Assistant for Domestic Affairs John Ehrlichman. According to Ehrlichman, when Nixon saw the size of the Earth Day crowd on the National Mall, he asked what he could do to become a "player" on environment. The aide suggested that they create an Environmental Protection Agency by executive order, under the rubric of government streamlining and efficiency. While Ehrlichman's account may be apocryphal, the timing of the memo that began the process appears to bear it out.[56]

On April 29, 1970, presidential advisors gave Nixon recommendations for the establishment of a new agency with power over standard-setting, enforcement on pollution, and the conduct of research. It envisioned bringing together numerous disjointed parts of the executive branch, each of which played some role in environmental affairs. This consolidation would end the long-standing separation between the Department of Agriculture and the FDA on pesticide regulation and bring parts of the departments of Health, Education, and Welfare (HEW) and the Interior involved in air and water pollution together under one new roof. Solid waste pollution was also included, and the memo proposed that some forty-four agencies would be brought together in what would become a new Environmental Protection Agency—the EPA.[57]

The next day, Nixon reversed his campaign promises to end the war and announced the expansion of US military operations into neighboring Cambodia. This led to strikes and demonstrations on sixty college campuses, culminating in National Guard units killing students at Kent State University in Ohio and police killing students at Jackson State University in Mississippi.[58] Clearly, the era of political chaos and disruption was not over.

Meanwhile, Muskie and his staff redoubled their already frenetic work to seize the momentum and push through a Clean Air Act Amendment (that would attack the problem at the federal level). The senator insisted that the Amendment be based on protection of the public's health, with science used to define binding requirements needed to safeguard it. Given the overwhelming public support for clean air, everyone in Congress, on both sides of the aisle, wanted to be part of the action. One Republican congressman frequently remarked, "I need to check that with the garden clubs back home," once again reflecting the influence of women in the environmental movement. And then he would nearly always side with Democrat Muskie.[59]

The 1967 Act had put the Department of Health, Education, and Welfare (HEW) in charge of setting auto emission standards. This cleaned up what came out of cars somewhat, but left it to localities to resolve the question of just how dirty the air could be when all pollution sources were considered. The earlier Act also gave HEW the job of setting up air-quality regions, but left the definition of allowable standards for concentrations of air pollutants and enforcement to the states within those regions, and it was the concentrations (or standards) that mattered for the air Americans breathed every day. This regional approach was a dismal failure, with less than a third of the regions having defined standards, and even fewer exercising any real enforcement.[60]

In June of 1970, the House of Representatives tried to claim the initiative by passing their version of an air pollution bill. They argued that the only way to fix the failures of the 1967 Act was to set air pollution concentration standards at the national level. This was an idea featured in Nixon's proposals, and Muskie and his staff feared that the Republicans would use national standards to appease business by aiming low, thus squandering the magic moment of popular support, when stringent changes in pollution levels could actually occur.

On July 9, 1970, the president sent a special message to the House and Senate about his plan to establish the EPA, but Muskie was wary. He also saw a threat in a bill that Senator Nelson had already introduced to ban the internal combustion engine altogether. The danger was that Nelson would tack his radical proposal onto Muskie's bill, thus killing the measure by overreaching.[61]

The tension between state and federal control is a central theme in American history, as well as in environmental regulation. The Constitution reserves all unspecified powers to the states, but automobiles and air pollution weren't on the framer's agenda in 1787. However, they empowered the federal government to regulate issues that cross state lines, and pollution from California was clearly drifting into Nevada. If the rules were not consolidated nationally, a race to the bottom might damage public health, with some states opting for very low standards to try to attract industries while exporting the resulting damage to their neighbors.

This is the point at which the battle for clean air entered a new phase. The widespread perception of a personal stake had spawned massive social action, and government had responded to the public will. But now it was time to engage that other powerful force in American life—the industrial corporation. While some businesses probably were overjoyed by this prospect, others with a nationwide customer base, such as the manufacturers of automobile and power plants, feared losing the markets in lucrative antipollution states like California and New York unless their products met the most restrictive rules. Churning out an array of different product lines for individual states would be prohibitively expensive. Besides, given the needs for investment and planning, industry generally prefers certainty to uncertainty, and regulation at the federal level would at least present a clear and consistent challenge. It also would provide a more compact target for lobbying, which could work to industry's advantage.

The first US national air pollution control act had been passed in 1955, but it did little more than provide federal funds for research. But then the Big Apple experienced another killer smog event that led to the deaths of two hundred people. Just when New York State was about to impose even tougher standards than California's, the federal government amended the 1963 Act to preempt additional states from adopting their own standards, cutting off the possibility of interstate races to the top.

Congress exempted any state from the prohibition against setting their own standards if they had already done so before March 30, 1966, and the only state in that category was California. US House member John Dingell represented the Michigan district that included the Ford Motor Company, and he tried vigorously to stop the California exemption. But the California delegation fought back, and could cite strong constituent support, along with states' rights and their established state law, as their rationale. An anti-Dingell radio program helped to build popular support for their side, and the exemption prevailed.[62]

California regulators didn't waste much time using their authority. In 1968, Haagen-Smit became chairman of California's newly established Air Resources Board. In 1969, when the Volkswagen corporation failed to certify its compliance with board requirements, Haagen-Smit banned the sale of Volkswagens in the state. Within ten days Volkswagen backed down and produced a compliance certificate.[63]

But in 1970, Haagen-Smit wrote, "There are ominous signs that control technology alone is not able to cope with the ever-increasing growth in population and all its polluting activities. Plans have to be made now as to how we want the basin to look and what kind of air we want to breathe a few decades ahead."[64] California and its Air Resources Board became a national "super-regulator" by using the California exemption to set higher pollution control requirements that gradually raised the national standards, decade after decade.

Finding a comprehensive solution to a nationwide problem would require more than skilled advocacy and political maneuvering. Large-scale solutions have to be practical, which requires synthesis and compromise. Ultimately, Ed Muskie and his committee took the process of technology steering on smog control to new heights by a legal mandate that forced the industry to develop technology that didn't yet exist, a strategy known as technology-forcing.[65]

Muskie's draft of the Clean Air Act of 1970 explicitly required the auto industry to reduce emissions of smog-producing carbon monoxide, organic molecules, and nitrogen oxides in new cars by a startling 90%, clearly a breathtaking number that would require an engineering breakthrough. To make matters even more challenging, manufacturers would have to reach those emissions targets in six years or less.

The Senate subcommittee held a press conference to release their bill on the eve of the August 1970 recess, and then they went on vacation. Meanwhile, as one staff member put it, "the auto companies went ape." General Motors, Chrysler, American Motors, and Ford all sent high-level executives to Washington to lobby against the draft bill, protesting that its standards couldn't be met. But Muskie didn't budge.[66]

After recess, the bill went to the full Senate committee. Leaders of the auto industry came to Washington again to complain that they were being asked to do the impossible. But this time they made the mistake of bringing technical people along with them. When one of the committee's staff stepped out of the meeting for a moment, a General Motors engineer followed him and, as often happened in that era, key information was communicated in the men's restroom. The engineer confided that "we can build whatever you tell us to build. If you tell us to build a clean car, we will build a clean car."[67]

The full Senate passed the Clean Air Act Amendment on September 22, 1970, by a vote of 73 to zero.[68] Muskie and his team moved on to negotiations with the House to arrive at the final bill that would go to the president. The House first agreed, but then welched on the auto emissions standards, which meant that it was time for an even harder game of hard ball. Someone in Muskie's subcommittee tipped off the influential columnist Jack Anderson that a House conference member's family owned a huge auto dealership, which represented a clear conflict of interest. Anderson ran with the scoop, which immediately spread throughout American news media, effectively neutralizing one of Muskie's key opponents.[69]

Nixon's proposals for an EPA had been moving along in tandem, with public hearings over the summer, which culminated in congressional approval in September. Nixon signed the executive order establishing the new agency on December 2, 1970. Meanwhile, the House-Senate conference finalized the Clean Air Act Amendment, essentially unchanged from the Senate version, and sent it to Nixon for his signature. On December 31, 1970, at the signing ceremony for what had come to be known everywhere— except in the White House—as "the Muskie bill," its namesake was conspicuously absent.[70] When queried by reporters, Nixon's deputy press secretary said that the bill had over forty sponsors and that the White House was unable to invite them all (although four Republican members of Congress made the guest list). Muskie was unfazed by the snub, commenting in his own press release on the same day that "although opposed in part by the Administration, the Act was a nonpartisan Congressional effort."[71]

A restless environmental champion, Muskie next trained his attention on passage of the Clean Water Act. In 1972, after 33 days of hearings, 171 witnesses, 470 statements, 6400 pages of testimony, and 45 subcommittee and full-committee markup sessions, the bill became law, completing a tour de force of environmental regulation.[72]

While the Clean Air Act itself had established nationwide emissions standards for cars, one of EPA's first jobs was to set national ambient air-quality standards. The law required that standards build in a margin of safety while protecting not just the average American, but also vulnerable groups, including the old, the very young, and the sick. By using scientific information about health impacts to define limits for the concentration of ozone, they arrived at a single number from coast to coast, meaning that all major components of smog—hydrocarbons, carbon monoxide, nitrogen oxide, and sulfur dioxide—from the industrial Northeast to the LA Basin, had to be reduced. So, along with auto emissions, this affected power plants and factories as well.

Industry was furious, and the godfather of the EPA himself, Richard Nixon, was not pleased. According to the EPA's first administrator, Bill Ruckelshaus, Nixon believed that the environmental movement "reflected weaknesses in the American character." He also told Ruckleshaus to control the "crazies" in his agency, but Ruckelshaus ignored the pressure and did what he was required to do by the act, hewing to what was scientifically defensible.[73]

Shrewdly, Ruckelshaus positioned the EPA as the guardian of every citizen's health, admitting that the standards were very tough but remaining stalwart about the need to protect the public. Pointing to the headwinds he faced, he added that "some American industrialists . . . misjudged the power of the environmental movement and its ability to galvanize public support. . . . We couldn't have invented any better antagonist for the purpose of showing that this was serious business."[74]

While the Clean Air Act mandated reductions in emissions from new autos and power plants, much more would be needed in some locations to make progress quickly enough to meet the standards required. States proposed such measures as changes in local public transportation, car-pool lanes in busy cities, increased regulation of existing power plants and factories, and more.

The Act preserved a measure of states' rights and precedents, and it sweetened the deal by providing federal funds for research and planning in city and state agencies. But under the new law, the federal EPA would still review each state's plans and progress. If a state failed to develop its own implementation plan by the federal deadlines, the EPA would summarily hand them one, providing even the most recalcitrant states with plenty of incentive to come up with their own plan.[75] Still, car companies had ample incentives to obstruct—one estimate suggested that each year the standards were pushed back could save the American auto industry about $5 billion in annual equipment costs. On the other side, challenges also came from environmental organizations like the Natural Resources Defense Council, which sought even tougher laws.[76]

The EPA also had the responsibility of establishing a new Federal Test Procedure for estimating vehicle emissions and certifying each car model. This altered test procedure meant that the automakers needed to reduce their emissions of carbon monoxide and hydrocarbons by about 25% by 1972 or face a fine of $10,000 per vehicle sold that couldn't be certified. In an era when a typical new car cost about $4000, this was a serious penalty. Car companies responded with minor modifications, which allowed them to meet the initially modest reductions required for 1972, but fuel economy and car performance suffered.[77]

An "information asymmetry" often exists between regulators and manufacturers, the assumption being that industry has a clearer picture of what's possible. This enables companies to pretend to be far less capable of innovation than they really are. So, erasing that asymmetry was central to the EPA's regulatory task. With more engineering data at their command, the agency could drive companies to spend more on serious R&D, which would lead to competition to come up with a better, cheaper solution to the emissions problem, putting teeth into technology-forcing. Doing so would require the EPA to develop some technology muscle of its own.

The EPA set up its own laboratories and began testing a wide range of options, particularly catalytic converters, long used to control pollution in manufacturing. These devices worked by passing exhaust gases over a metal surface—the catalyst—that converted the pollutants into compounds that produced less air pollution. In factories, though, they were stationary, while conditions in a moving automobile were constantly changing. Nevertheless, by equipping their test vehicles with converters, the EPA showed that these devices could meet the standards.[78]

GM and Ford were on board with the idea of catalytic converters, but that meant gasoline had to change, too, because the lead that had been added for decades to improve performance would destroy the catalyst. Muskie had tried hard but failed to change the rules based simply on the considerable risks of leaded gasoline for human health. However, the wily senator now realized that the enormous public support for action on smog gave him a political lever not to be wasted. Could it be that the rapid and aggressive Clean Air Act emissions schedule was set to ensure that the lead-averse catalytic converter was the *only* way to meet the timetables, thereby forcing out lead and killing two of America's pollution perils with just one stone? Interestingly, the Act already had given EPA the explicit authority to regulate the composition of gasoline, meaning that it had all the tools it needed to make such a roundabout plan work.

Muskie had learned in earlier Senate battles that having one expert debate another was not the way to defend the science underlying a policy change. That's why the Clean Air Act contained explicit provisions that scientific and technical issues would be evaluated in reports carried out by the National Academy of Sciences. A consensus among experts by such a respected organization would build a strong factual foundation from which EPA could work. The first academy report in 1972 suggested that the costs of compliance with the Clean Air Act would be less than half of that estimated by industry, but even the academy was doubtful that the catalytic converter could be practical by 1975.[79]

Volvo promptly requested a delay of a year, followed by Detroit's Big Three. EPA rejected the request, and the issue went to the courts, which remanded it back to the EPA. In 1973, EPA granted a one-year delay for the 90% carbon monoxide and hydrocarbon emission reductions, but set an interim reduction level of 50%. Then California flexed the special muscle allowed them by the law to set their own, even lower standard—a 75% reduction—for cars sold within the state in 1975.[80] Most other companies continued struggling to meet the California law with existing technologies, but the Swedish company Volvo did an about face when confronted with loss of the lucrative California market and became the first to install catalytic converters. This essentially forced US automakers to spend on the R&D needed to develop and implement the devices.[81]

American car companies achieved substantial progress with this first-generation catalytic converter, but reaching the mandated reductions in nitrogen oxides would take a more complicated second-generation system. These standards were delayed even more, because everyone knew that the necessary technology was much further away. Then in late 1973, the world was rocked by a war between Israel and its Arab neighbors, which led to an Arab oil embargo against nations that supported Israel, including the United States. The price of oil shot up, and the contentious issue of whether auto emissions reductions would hurt fuel economy became critical to the debate. In mid-1974, Congress extended the interim standards by another several years.

Even so, more than 80% of American cars produced in 1975 sported catalytic converters that reduced carbon monoxide and hydrocarbon emissions by at least 50%, and automakers had acquired the technology and experience needed to reduce emissions still further. Moreover, unleaded gas became available to power them, growing in market share year by year as more new vehicles came off the assembly line.

While carbon monoxide and hydrocarbon emissions were dropping, the 1975 standard for nitrogen oxide still stood at 3.1 grams per mile, far above the Clean Air Act target of 0.4 for 1977. Reaching the goals for nitrogen oxide would require onboard computer control. That meant that the next challenge facing implementation of the Clean Air Act was to develop computers able to function in the heat and vibration of a car engine. Competitive pressure spurred industry in Asia, Europe, and North America to begin developing these onboard systems.[82]

One way to measure the success of technology-forcing is to examine the number of patents granted during years in which legislation puts pressure on industry to innovate. Between 1968 and 1972, the number of patents issued each year for automobile emission control technology in the United States soared by more than a factor of ten. Between 1974 and 1977, as deadlines for emissions standards were pushed back, the number of patents fell by about half. By 1977, Japan had surged to become the world's global emissions control leader, with roughly twice the number of patents as the United States per year.[83]

The cars coming off the assembly lines in 1978 did not meet the required standards, and the EPA had already granted all the delays the law allowed. Because these cars could not be sold unless certified, Congress had to amend the Clean Air Act or shut down the entire US auto industry. Muskie, concerned that Michigan's senators would come in with a new amendment dictated by the Detroit car makers, struck a deal with his longtime Republican ally, Senator Howard Baker, who wanted to relax the nitrogen oxide requirement from 0.4 grams per mile to 1.0. Since the auto manufacturers wanted much more than this, Muskie promised Baker he'd agree to 1.0 and work to make sure that this compromise passed the Senate. When Michigan Senators Robert P. Griffin and Donald W. Riegle Jr. jointly sponsored a much more industry-friendly amendment, Baker promptly offered his as planned, and, naturally, Baker's more reasonable amendment was adopted. Along with relaxing the standard, the deadline for compliance was set back to 1981, giving the industry much-needed breathing room.

But the next step was the conference with the House, where Representative John Dingell of Michigan sponsored a bill similar to the defeated Griffin-Riegle proposal. Despite Dingell's push, Muskie stood pat, saying, "There aren't any auto plants in Maine. If you want to close down the auto companies, then that's your decision." Dingell stormed out of the room and ultimately let the politically unstoppable Senate bill prevail, perhaps because he had promised industry supporters more than he could deliver and felt it would look better to the folks back home if he wasn't even there.[84]

By 1981, all new vehicles sold in the United States met the new standards. This was a huge victory, long in coming, but it was the end of only the first battle, not the war. A decade later, California used its special powers again by requiring that 2% of all cars and light trucks sold in the state by 1998 be zero-emission vehicles, meaning all-electric. The technology to do this at reasonable cost failed to materialize, and the state ultimately pulled back. But here again, some studies indicate that the mandate served as technology steering to advance another approach: the hybrid.[85]

Carmakers in the United States, Japan, and France had let R&D on batteries languish for years, but responded to the California electric vehicle requirement by greatly increasing investment in research. Within a few years, a boom in advanced battery technology allowed rapid development of the hybrid auto, which did not eliminate emissions, but typically cut them by a factor of two.[86]

The affordable electric car made a comeback in the 2010s as technology advanced. Still, LA has the worst air in the nation, and it has yet to fully meet required standards. In 2020, COVID-19 offered an unexpected window on the problem. During the pandemic, traffic in LA County essentially shut down. With near zero emissions at all from gasoline-powered vehicles, LA's air cleaned up within days and stayed that way throughout the worst of the health crisis.[87] This showed what the regulators and experts had long known: the air in the LA Basin probably will never fully meet national air-quality standards until nearly all its vehicles are electric.[88]

For all its effects on the auto industry, the Clean Air Act of 1970 wasn't just about cars. It also put pressure on power plants to clean up their emissions of sulfur dioxide and particulates. When Muskie designed the Act's provisions, he was well aware that coal-fired power plants were the source of more than half the nation's sulfur dioxide pollution. People living near coal plants routinely suffered from the rotten-egg odor of sulfur gases, and the sulfur dioxide formed a choking haze of small particles, along with sulfuric acid rain downwind as well as nearby. Research would later demonstrate that the particles generated by the sulfur not only obscured the skies, but also increased human respiratory illness and deaths from heart attacks and cancer. To meet the federal air-quality standard for sulfur dioxide, plants would have to install expensive scrubbers or burn low-sulfur coal. The EPA again attacked the information asymmetry by demonstrating feasibility in their own laboratories, and Japanese innovations continued to help spur further advances. Early efforts at making scrubbers were plagued with frequent breakdowns and inefficient operation, but by 1977 the new models worked 90% of the time.[89]

While the Clean Air Act of 1970 achieved significant reductions in sulfur dioxide emissions per kilowatt hour of energy generated, rapid increases in demand for energy, driven by increasing population, led to widespread increases in sulfur pollution. Due to both cars and power plants, nitrogen oxide pollution increased. These gases turn into sulfuric acid and nitric acid, which then falls to the ground in the form of acid rain.

The basic chemistry of this form of pollution has been known since at least 1872, when Scottish chemist Robert Angus Smith explicated it, but the public began to express concern only in the mid-1980s, when it became widely perceptible that forests and lakes were being damaged in the United States and United Kingdom. Less industrialized neighbors, such as Canada and Sweden, suffered as well, which made this environmental threat an international issue.[90]

In the northeastern United States, trees turned brown, and formerly abundant fish in lakes and streams died off. The faces of marble statues that had been in place for a century or more became pockmarked and began to crumble. But this was the era of Ronald Reagan, whose concerted attack on the federal government's institutions, powers, and regulations had gutted environmental protections, which meant that policymakers and advocates, not just engineers, were going to have to innovate.

Reagan assumed office in 1982, and within a year, the number of enforcement cases referred to the Department of Justice by the federal EPA had fallen by about 80%.[91] But public support for environmental controls revived by the end of the decade.[92] Reagan's vice president, George H. W. Bush, campaigned on a promise of being the "environmental president,"[93] and when he came to the White House in January of 1989, the time was ripe for new efforts to improve America's air, partly because Americans were, once again, able to perceive the rise in pollution in more personal terms.

In March of that year, another massive oil spill—the *Exxon Valdez* disaster—fouled pristine waters off the Alaskan coast and put the pitiful sight of oil-drenched seabirds back onto the evening news. That created empathy, but the largest injection of "perceptible" into air and water pollution was the toxic rain spoiling the public's enjoyment of the natural world in their waterways, parks, and open spaces.

After years of Reagan's antiregulation rhetoric, however, the president of the Environmental Defense Fund activists group, Fred Krupp, reached out to a White House staffer to discuss a "market-based" solution more in keeping with Republican economic thinking. Krupp proposed setting up a trading system aimed at cutting the pollution in half. Power companies would buy permits allowing them to emit a capped amount of pollution, but they also would have the right to bank or sell those permits to other companies if they wished. This "cap-and-trade" market would provide a profit incentive for doing more than the minimum to limit their own pollution. But cost effectiveness and cost-benefit trade-offs are notoriously hard to evaluate when they involve value judgments, such as the delta between dollars saved by industry versus, in the case of acid rain, the non-quantifiable costs of diminished public enjoyment of forests, lakes, and streams. Other environmental groups were skeptical about the proposal, mainly because this system of permits effectively *authorized* pollution. Even so, Krupp persuaded the skeptics to go along with this pragmatic approach—because it might actually work.[94]

A draft bill was crafted by Environmental Defense Fund, White House, and EPA personnel, but the gnarly problem of how to check power plants for compliance persisted. How heavy-handed could EPA be? Would periodic inspections be sufficient, when the companies could just turn off scrubbers or change fuels when the regulators weren't looking? An EPA policy analyst came up with the idea of putting continuous measuring devices on the stacks of every power plant. This way, instead of complicated inspection procedures, all you had to do was measure what was being emitted and, at the end of every year, compare the readings to each plant's allowed level of emissions. This took the regulators out of the loop, displeasing them, but providing an option expected to appeal to the public. President Bush supported the bill, and even sought to lower the cap to the level Krupp was promoting. Eventually, the House and Senate adopted the White House level.[95]

"Authorizing" pollution may have sounded heretical to many environmentalists, but the proof was in the pudding. Between 1990 and 2004, sulfur dioxide emissions from power plants decreased from 15.9 million tons to 10.2 million, substantially reducing acid rain, despite a 25% increase in electricity generation during that period. By 2010, emissions had fallen further, to 5.1 million tons.[96]

Conservative economists often hail the Clean Air Act amendments of 1990 as a demonstration of the triumph of the market. Acid rain did rapidly decrease on average as emissions dropped, but the legacy of changes in soil composition in some areas means that certain bodies of water, including some granitic lakes in the Adirondacks, will not recover for centuries. Although the acid rain quickly stopped falling as emissions dropped, this shows another avenue for persistence of environmental problems: ecosystem recovery times. These differ by soil type, so that many other waterways and wild areas have substantially recovered from acid rain, and forest die-back—with trees dying in droves from it—is now mostly a thing of the past.[97]

A purely fortuitous event played a role in reducing the use of high-sulfur eastern coal, which may have made the "cap-and-trade" numbers look better than they would have otherwise. The deregulation push of the Reagan administration reached America's railroad shipping costs just as cap and trade began. Deregulation meant that relatively clean low-sulfur coal from Wyoming's Powder River Basin could be shipped at a lower price per pound per mile, and this unanticipated and unrelated factor made the cap-and-trade system appear more effective than it would have been without it.[98] The benefit varied depending on distance to the power plant, which meant that the price advantage accrued mostly to plants in Michigan and Indiana, and in an arc down through southern Illinois and Missouri.[99]

The Clean Air Act probably serves as the most spectacular example of why pollution control is almost always a wise investment, no matter how "market oriented" your economic philosophy. By some estimates, the Clean Air Act has yielded benefits in the several trillions, largely from avoiding the costs of both mortality and health care. Remarkably, a large part of this occurred because human health is affected even more by small particles formed in air pollution than by the gases, a factor that was not known at the time of the Act's design. When all is said and done, from 1970 to 1990 alone, the Act is estimated to have saved America somewhere between 5.6 and 50 trillion 1990 dollars, which corresponds to about 10 to 100 trillion in 2020 dollars.

The net benefits from 1990–2000 of the tighter sulfur dioxide cap-and-trade program are estimated at about 58 to 114 billion in year 2000 dollars,[100] or about 90 to 170 billion in 2020 dollars. The Act was ultimately among America's most practical pollution control measures.

The United States was a world leader in cleaning up air pollution, reflecting a strong confluence of the personal, perceptible, and practical. Between 1970 and 1976, American auto emissions dropped substantially faster than those in Sweden, a country long considered one of the world's most environmentally minded. But the provisions of the Clean Air Act also benefited people everywhere, because the technologies developed to reach emissions goals became cheaper for everyone over time.[101]

However, analysis shows that the legacy of "redlining"—the unfair distribution of federally sponsored mortgages in nonwhite neighborhoods that began in the 1930s—led to increased air pollution in those areas due, for example, to the siting of freeways, heavy industry, and power plants.[102] These polluting infrastructure sources have a long lifetime of many decades. Thus, though smog itself is short-lived, they will continue to spew out unhealthy substances in communities of color, irrespective of income level, until major infrastructure changes take place. Because non-Hispanic black communities continue to suffer poorer air quality than other demographics, even in the twenty-first century,[103] public protests for environmental justice include a focus on air pollution.

Worldwide, many urban areas in poor countries have grown into megacities, often with dramatic increases in pollution. With the remediation of poverty and disease seen as being more urgent, even fairly cheap pollution control is considered a luxury. But as lower-income countries become better able to meet their basic needs, people are already demanding cleaner air, just as Americans did. When Mexico City began work to clean up its smog problem in the early 1990s, its citizens and policymakers were quickly successful because technology was already available to do it, demonstrating once again how lower-income countries can benefit when richer countries bear the costs of technology development first.[104] China and India both have notorious air pollution problems, and their cities have vied for the dubious title of world's worst polluter in the 2000s. But as their economies have grown, so too has popular support for cleaning up the air. Chinese air pollution control measures have notably increased, while India's have been slower to be effective.

It seems hard to imagine that any law could do more for people and the planet than the Clean Air Act did by effectively combating not only smog but also acid rain. In an indirect way, the Act would also prove effective in stopping something else that had been imperceptibly poisoning children the world over for centuries. And, as in the case of smog, social activism and science were pillars that combined with policy to make that happen.

4

Lead

A METAL'S HEAVY PRICE

In 1969, two-year-old Gregory Franklin emerged from a coma with permanent, severe brain damage. The cause: lead poisoning. The Franklin family told the media how they begged and fought with their landlord to remove the peeling lead paint from their apartment for a year and a half before Gregory was rushed to the hospital. After Gregory's sister also tested positive for lead, a local group of concerned citizens formed an action campaign to test the paint in the Franklin home and provide that data to city officials. But still the landlord and the New York City Health department did nothing. The mayor and his associates already knew full well that the Franklins lived in what city hall privately called the "lead belt," the thousands of tenements stretching from Harlem to the Bronx that had nominally been declared unfit for occupancy but remained home to many families of color and recent immigrants.[1]

One unlikely group realized that it was going to take more than data from one home to move the needle on lead poisoning and began a determined push to ensure that it happened. They sported the purple beret and military camouflage uniform of the Young Lords, a street gang that had transformed itself from violence and drug-dealing to militant civil rights activism. The Young Lords were Puerto Rican nationalists, and their flag included an ominous image of a raised arm holding a rifle. But the bravado of their flag belied a profound pivot by the gang to constructive work to advance civil rights.[2]

The struggle for civil rights in America pre-dated the era of the American environmental and anti-Vietnam War activism of the late 1960s and 1970s, but arguably contributed to their foundations. Civil rights activists increasingly questioned the status quo during the fifties and sixties, and they learned what protest methods could work. The Young Lord's tactics included a nonviolent sit-in at the city's department of health that not only drew widespread attention from major media outlets but also led to the production of hundreds of lead-poisoning test kits. Together with students from New York Medical College, they carried out tests door to door in the affected areas.[3] When fully 30% of the children tested positive for lead, they had ample evidence to frame the issue of lead poisoning in a new light: racial justice. As the coverage of their activities grew in an era of attention focused on change and the systematic questioning of all authorities, the city finally transformed the housing laws to require landlords to remove lead hazards or pay for the emergency team that the city would send within five days of inaction.[4]

With so much public attention brought to bear on lead poisoning through the efforts of the Young Lords and dozens of other civil rights organizations, important steps were finally taken nationally in the late 1960s, including greatly expanded blood screening programs for children. In a surprising twist, that data revealed that lead poisoning was widespread across the country, not just in poor black neighborhoods nor only in urban areas but everywhere and for everyone—white, black, rich, poor, urban, rural, and more. To be sure, the highest rates of excess lead in children's blood were found in the dilapidated buildings where so many urban minorities suffered with peeling lead paint, but rates elsewhere were shocking too. Civil rights inspired the data that revealed the scope of childhood lead poisoning.[5] The outcome of that testing transformed the issue into a deeply personal one for every American parent and began the process of better management of an environmental issue that dated back not just decades, but millennia.

In 1921, General Motors announced that a team of scientists, including Thomas Midgley, Charles Kettering, and Thomas Boyd, discovered that tetraethyl lead added to gasoline would allow the development of much more powerful, high-compression auto engines. The additive increased octane rating and silenced the engine "knock" that happened when fuel ignited too early and unevenly in the compression cycle. Knock had plagued the industry for years in their quest to make the automobile a faster and more powerful conveyance.[6] The oil industry and chemical companies jointly embraced tetraethyl lead, as a new and highly marketable product.

The potential boon to motoring soon suffered a serious and sad setback, when over a dozen workers making the chemical at an American plant died in 1924. Many more were sickened. Midgley asserted that the workers had caused their own problems by ignoring warnings and failing to protect their hands and arms. The US Surgeon General suspended the production and sale of tetraethyl lead, and ordered a scientific panel to carry out a safety study. The panel deliberated for a time but found that there were "no good grounds for prohibiting the use of ethyl gasoline . . . as a motor fuel, provided that its distribution and use are controlled by proper regulations." They blamed careless workers, lax industrial processes, and inadequate regulations instead of the chemical itself, and production promptly resumed.[7] While deaths in an occupational setting like the chemical plant were obvious, no one considered the possibility of less obvious and much broader exposure slowly and subtly causing chronic damage to the population at large, so lead poisoning was not yet a personal issue outside the workplace. Tetraethyl lead additives to gas would continue to be sold for much of the twentieth century, adding lead to the air and soil across America and elsewhere in inexorably growing amounts. For marketing purposes, the material was dubbed "ethyl," dodging any negative connotation that consumers might have associated with lead.

Anyone with a sense of history should have anticipated the lead poisonings at the American chemical plant because they had been preceded by so many others. Lead poisoning had sickened or killed people not just in the twentieth century, but over thousands of years. Lead's softness hints that it will melt at lower temperatures than many other metals, and is therefore relatively easy to extract and work. Early decorative lead statues have been dated to around 7000–8000 BCE. Lead came into broader use around 3000 BCE in finishes on Egyptian pottery. Such a practice on statues might pose an occupational risk just to the workers who made them, but its use on dishes and cookware introduced it into the bodies of early Egyptian consumers.[8]

Centuries later, the Roman need for materials for aqueducts to provide homes and villages with water ushered in an era of lead mining on a much larger scale. As they worked the lead, vapors and particles spread over the planet, like other forms of pollution. The legacy of Roman lead mining spread around the hemisphere, even reaching Greenland, where ice-core data retrieved almost 2000 years later would one day help awaken the world to the enormity of the lead risk.[9]

The Romans also used lead to make cups, plates, and cooking vessels. Cooking in a pure copper pot imparts an unpleasant taste when grapes are boiled down to the syrup used to make some wines. Lead pots have the unfortunate quality of imparting a sweet taste due to formation of lead acetate. The Roman scholar Pliny was among those who advocated that "preference should be given to lead vessels in boiling defrutum (grape syrup)." Lead-sweetened wine was widely consumed by upper-class Romans. Modern scholars have continued to debate whether or not these dangerous practices contributed significantly to the end of the empire, but reports of lead poisoning aren't hard to find in Roman writings. Roman author and engineer Vitruvius was one of the first to suspect that the use of lead posed a health risk, writing in 8 CE that "When gold, silver, iron, copper and lead and the like are mined, abundant springs are found, but mostly impure . . . When the water is taken into the body . . . the muscles swelling with expansion are contracted in length. In this way men suffer from cramps or gout . . ." He also noted that "water is much more wholesome from earthenware than from lead pipes."[10]

Long after Rome, lead continued to enjoy wide usage worldwide despite its risks, including in lead water pipes, foods, and even cosmetics. It was not until the 1400s that Spain, France, and Germany recognized the dangers of using lead-sweetened grape syrups in wine and banned them. Lead was used not just in wines but also in presses employed to crush apples for the famous cider of southern England, enhancing the sweetness. In the 1700s, it caused the widespread malady commonly called the "Devonshire colic,"[11] every autumn during apple harvest, characterized by "tumults in the bowels," pain, seizures, and occasionally death.[12] Nonetheless, the versatile metal was used by the canning industry as a solder in many food cans, and it was not until the 1990s that the US Food and Drug Administration banned all domestic and foreign cans soldered with significant lead to keep it out of the food supply and the bodies of the American public.[13]

But of all the many and varied sources of lead, just two have dominated the widespread human suffering it has inflicted in the industrial era: gasoline and paint. The durable, bright white finish that lead imparted to paint made it a popular choice. But as early as 1848, L. Tanquerel des Planches noted that children could develop "colic" from putting toys painted with lead into their mouths.[14] At a 1923 conference in New York, an American professor described a child's poisoning by gnawing paint from his crib, as well as his studies of "dementia" induced by lead in guinea pig experiments.[15]

Australian physician J. Lockhart Gibson's 1904 article "A Plea for Painted Railing and Painted Walls of Rooms as the Source of Lead Poisoning Amongst Queensland Children," revealed that not just painted toys and cribs, but the very walls of the typical home posed a deadly risk to young children. Gibson noticed that lead paint loses its oil and gloss to become a dangerously powdery surface and source of hazardous dust. He stressed that its stickiness meant that the dust adhered to "fingers and nails by which it is carried to the mouths of children, especially in the case of those who bite their nails, suck their fingers or eat with unwashed hands."[16]

Why wasn't action taken to stop using lead paint? Despite a growing body of medical literature documenting the dangers of lead, the risks to the consumer remained little discussed.

Again, it was easy to blame the victims—the children and the parents—by arguing that the normal habits of mouthing displayed by children are somehow improper and would of course be stopped by any responsible parent. The lead industry in the United States and elsewhere deftly used this argument to avoid any restrictions on their products for the better part of a century, even after Lockhart's work.

Other researchers were broadening the lead-risk landscape to extend beyond paint. In the first two decades of the twentieth century, pioneering industrial toxicologist and physician Alice Hamilton documented the high mortality of workers in American lead and enamelware companies, as well as in rubber production, painting, and batteries. Hamilton became renowned as the leading US authority on industrial lead. Her research led to her to suspect that much of the lead poisoning she encountered occurred through lead dust and fumes in the workplace. While she campaigned vigorously for increased cleanliness (sometimes through such simple measures as allowing workers the time and opportunity to wash their hands before lunch), she was also well aware of the larger dangers, writing, "It is not so easy to understand why we have so long been in ignorance, . . . why American physicians and sanitarians, to whom all other questions of preventable disease are matters of the greatest interest, should for so long have neglected industrial plumbism."[17] Hamilton pointed out that workers carried lead dust from the workplace into their homes, resulting in risks both to themselves and their babies. After the poisonings at the American tetraethyl lead factory, she was outspoken as well as prescient in her criticism: "I am not one of those who believe that the use of this leaded gasoline can ever be made safe. . . . Where there is lead, some case of lead poisoning sooner or later develops."[18] Hamilton became respected enough in the field of industrial toxicology to become, in 1919, the first female faculty member at Harvard University—although her appointment expressly forbade her from entering the faculty club or receiving faculty football tickets.[19]

In the wake of a tragic world war, the International Labour Organization was formed in 1919, the first specialized agency of the League of Nations. Its mission was to navigate the views of workers, employers, and governments in shaping responsible labor standards and policies of global importance. Among the organization's first actions was the 1921 convention on white lead in painting, sparked by the high incidence of sickness and death not in consumers but in the painters. Painters handled the toxic substance daily and faced the greatest risk, whether from the dust or tiny drops breathed in when painting, finishing, or sanding, or from handling the materials. Painters of the era were known to suffer high rates of lead poisoning. Although the dangers were personal and perceptible to the painters, the non-occupational risks were still unrecognized by many people, despite known cases of lead poisoning in children.

Discussions in the International Labour Organization were heated, and covered a gamut of issues including mechanisms of lead poisoning and diagnosis, statistics on the extent of risk to painters, and, perhaps above all, whether substitutes would work as well. And what effect would a ban have not just on lead mining, but also on the prices of interconnected metals like gold, silver, and zinc. Just when it seemed negotiations would break down, a diplomatic compromise was struck between those seeking a total ban and those arguing that no formal agreement was needed at all. The White Lead Convention of 1921 limited lead to less than 2% for interior paints only, and allowed for exceptions in railway stations and industrial establishments. It also included measures aimed at reducing dust and increasing cleanliness by the painters.[20]

As with international agreements today, the next step was ratification by member countries. As of the 2020s, more than sixty nations had ratified.[21] Like all international environmental conventions, the study of which nations signed and which did not reveals how diplomacy is shaped and how human tragedies are made. Despite J. Lockhart Gibson, Australia has not ratified to date, nor has the United States. Both were major producers of lead, along with Britain.

The British chose to regulate work with lead paint by requiring wet instead of dry sanding of surfaces, along with increased painter cleanliness and handwashing.[22] Using wet sandpaper sweeps more of the lead paint dust onto the paper, while dry sanding sprays it into the air. One more time, the blame fell on the victims, who just had to be a little more careful in their work.

Among the boosters and early ratifiers of the International Labour Organization White Lead Convention were France and Belgium. But there are very rarely any angels among nations involved in environmental negotiations, especially when valuable mineral assets such as metals, coal, and oil are involved. France and Belgium were the key producers of the rival and much safer pigment, zinc oxide. They had already established domestic bans on interior lead paint to boost their own zinc industries, so joining the agreement was an easy decision.[23]

The National Lead Company both mined the metal and produced the paint in the United States, using the Dutch Boy as their marketing face for most of the twentieth century (until its paint business was sold to the Sherwin-Williams Company). The smiling face of the Dutch Boy was prominent in the company's advertising, some of which explicitly targeted children. A promotional paint coloring book for boys and girls entitled "The Dutch Boy's Jingle Paint Book" was given to buyers, one of many advertising products used to convey the fun that painting with lead would provide for the little future consumers at whom the gift book was aimed. National Lead promoted the use of Dutch Boy lead paint with rhymes within the coloring book, one of which included this now-remarkable snippet:

> A paint that's always certain
> To stay where it is spread
> And save folks lots of worry
> Is Dutch Boy pure white lead.

Scientists and public health officials continued to express concern over lead in ensuing decades, but to little effect. The regulation of paints fell largely to local building codes or municipal regulatory mechanisms, and only a few chose to act. Baltimore was the first city to take the bold step of banning the use of lead paint in new housing in 1950. This useful measure stopped further spread, but did not address the widespread lead paint already present in countless homes in the city by then, and paint in old homes continued to poison Baltimore's children in the twenty-first century.[24] Only a few other localities followed suit.

The history of lead use in America highlights the way public health can fall into the cracks of government's structure. There were no US national mechanisms to deal with concerns like lead paint or leaded gas—no national laws, and no national agencies charged with overseeing environmental pollution until the latter part of the twentieth century. Pollution control was entirely local, and in many cities it simply focused on "dense smoke," utterly missing any unseen hazards not only in the air but also in consumer products. There was no national funding system for conducting the independent scientific studies that might help establish risks. So a vacuum occurred in the information pipeline, and it was skillfully filled by industry.

The Lead Industries Association formed in 1928 and included all of the major US companies mining and refining lead, as well as many of the users, such as paint manufacturers. Until the 1960s, the available US science on lead was dominated by laboratories funded by the Lead Industries Association, notably the Kettering Laboratory at the University of Cincinnati, named after one of Midgley's co-discoverers of tetrathyl lead's properties as a gasoline additive. Its leader was Robert Kehoe, a toxicologist and environmental health specialist who believed that low levels of lead in the body were natural. Kehoe frequently stated his certainty that the body could adapt to expel excess lead until it reached very high levels, and only recognized the hazards to children.[25] He reached this conclusion based in part on an experiment he conducted in Mexico, published in the prestigious *Journal of the American Medical Association* in 1935. Kehoe measured the lead content of the food consumed by nine healthy young indigenous men, as well as the lead content of their feces and urine, arguing that these subjects were not exposed to lead at work. He reported that the main source of the lead in these subjects was their food, and that the amount excreted was consistent with the amount consumed, within uncertainties. Interestingly, Kehoe found that the amount of lead in feces was actually slightly higher than the amount eaten during the twenty-four-hour period he studied, but he does not comment on this finding, except to say that it is within estimated uncertainties. He does not mention whether the dishes these men used were lead-glazed, as was much Mexican pottery of the era. Kehoe concluded that "apparently an equilibrium is reached after a time, so that a substantially constant concentration of lead remains in the tissues, and lead output becomes equivalent to lead intake." Whether twenty-four hours is an appropriate time interval to draw such a sweeping conclusion was not addressed, nor was the question of whether short studies of only nine adult subjects would be sufficient. Kehoe became the voice of science on this issue for decades, based largely on this single study.[26] Fortunately, a single study, even if great (and this one doesn't seem to me to be one of those great ones!) holds no such sway in the world of environmental science today. Typically verification or refutation by others happens quickly whenever researchers reveal important new findings. Scrutiny is essentially guaranteed by the diversity of research groups and funding sources, along with increased caution by policymakers, who require more studies to ensure their own credibility in reacting to concerns. But Kehoe was able to reign supreme and uncontested in his very important field of metal toxicology for decades.

It was an unlikely challenger who overturned Kehoe's interpretation of the risks of lead. Claire Patterson was a geochemist whose passion was to measure the age of the universe using isotopes, a long-standing brass ring of his chosen scientific field. In colloquial terms, he badly wanted to know exactly when the Big Bang—the origin of the universe—occurred. He employed a method similar to the carbon dating that can tell us about the age of fossilized bones of ancient peoples or animals, but using uranium allowed him to do so on far longer timescales. Patterson worked at Cal Tech as a postdoctoral associate in 1953, funded in part by the Atomic Energy Commission. Patterson's boss promoted this work to the commission on the basis that it was important to them to understand how much uranium could be extracted from rocks. No one told the commission that the scientists were also planning to research the age of the universe.[27]

Patterson and his boss also convinced oil companies to fund them for ocean-sediment analysis, since understanding ages of bands of material found in a drill core had the potential to help identify where oil deposits might be found. They paid extreme attention to keeping their samples and laboratory clean in their quest to measure the age of the universe. Analysis of the carefully collected ocean sediments revealed a surprising and disturbing characteristic: there was a mysteriously large amount of lead in the upper, newer portions of a sediment core, far more than in the deeper layers that had been laid down centuries earlier. The researchers only had data from a few remote sites in the Pacific and Atlantic, but all showed this curious feature.

Patterson made some rough estimates of where the lead in the shallow sediments might be coming from, and concluded, "If you took those profiles and you extrapolated from that over all the world's oceans, the amount of lead equaled what was being produced from gasoline. It could easily be accounted for by the amount of lead that was put into gasoline and burned and put in the atmosphere." When he revealed this finding, the oil companies immediately stopped funding him. So he pressed on, with the Atomic Energy Commission funds alone.[28]

Patterson knew what he needed to prove his hunch: "the snow that never melts in the polar regions." Even in the remote poles, lead would be in the air, and snowfall would incorporate it into the deep ice that covers both Greenland and Antarctica. He explained that "Lead is in the snowflakes. It goes down, and you have a layer there. Next year you have another one."[29] But getting to the top or bottom of the world wasn't easy in the 1950s.

The development of the next phase in Patterson's work meshed well with a major change in the way American science was conducted. No longer would industry enjoy its commanding monopoly on scientific inquiry: the National Science Foundation (NSF) was established in 1950 with a mandate to fund a broad range of academic research. Fortunately, Patterson's quest to obtain a polar snow core came at an especially propitious time: the recently formed NSF was just about to participate in an ambitious international endeavor to make geophysical measurements worldwide, the International Geophysical Year. The International Geophysical Year made exploration of the remote polar regions a hot research topic in 1957, just at the moment when Patterson approached the NSF for funding. He secured support and promptly flew to Greenland with several Cal Tech summer students, where they embarked on one of the most important and strenuous student research programs ever conducted. They began to dig. And dig, and dig. They cleaved massive amounts of snow out to form deep shafts. For each data point, they "had to dig shafts down—200 or 300 meters deep—to go back in time," Patterson said, estimating that digging to that depth would be the equivalent of looking at data from AD 1700. The backbreaking work of pulling out two-foot cubes of snow in the chilling cold paid off in an extraordinary result. The Greenland snow showed clearly the increase in lead during the twentieth century, some two to three hundred times the lead in the older snow samples, mirroring the behavior in the ocean sediments. Patterson had his proof.[30]

Patterson next spent years studying how lead and other elements are absorbed in the body, branching out into biological chemistry. When asked if his motivation was human health, Patterson passionately protested that his motivation was entirely geochemical: "I wanted to know, what is this natural level of lead? I didn't care two hoots about verifying what the contamination was. I was forced to measure the contamination in order to arrive at what was the natural level."[31] Yet in 1965 he published a landmark paper in an environmental health journal that contained an extensive analysis of lead inputs to the environment, including a revealing focus on the difficulty of finding a pristine place anywhere on the planet. He emphasized that if lead was in Greenland's ice cores, then lead was virtually everywhere. That was consistent with what he had guessed about its sources. He also concluded that this meant there were no longer any truly natural blood lead levels to be found in modern humans anywhere on Earth. He went on to state that this shocking result implied that the "typical" levels of lead found even in remote regions posed an enormous threat to human health through "severe chronic lead insult." In short, lead poisoning was taking place worldwide. He based his conclusions in part on an analysis of the ratios of concentrations of lead versus other metals, such as calcium and barium, in food and blood. Although much of the paper is written in the calm and impersonal tone of science, parts of it reveal an angry man with few inhibitions about political remarks. He was, for example, outspoken in his condemnation of the United States Health Service, finding their limited research in this area "conspicuous and disgusting" and concluding that "in common with other kinds of technological filth, [lead] may bring agony into our existence."[32]

Well before Senator Ed Muskie shepherded the landmark Clean Air Act of 1970 into being to combat smog, his environmental awareness led him to an interest in lead in gasoline. As chairman of the Senate Subcommittee on Air and Water Pollution, Muskie put the hazards of lead into the public eye by convening hearings in 1966. Among the witnesses to testify, the Surgeon General, William Stewart, acknowledged studies associating lead exposure with the occurrence of mental disabilities in children. The fireworks occurred when Kehoe and Patterson laid out their opposing views in the public forum that congressional hearings represent, even though they did not appear on the same day.

Kehoe asserted that his long research experience made him the world's uniquely qualified judge of the dangers of lead, flatly stating that "I would simply say that in developing information on this subject, I have had a greater responsibility than any other persons in this country. . . . The evidence at the present time is better than it has been at any time that this is not a present hazard." When asked if it would be desirable to find a substitute for lead in gasoline, Kehoe focused on his view that lead, as a "natural" substance, could not pose a threat, completely ignoring its potential toxicity and the enormous increase in the amount of lead that Patterson had documented using Greenland's ice: "There is no evidence that this has introduced a danger in the field of public health. . . . the work of the Kettering Laboratory in this field [shows] that lead is an inevitable element in the surface of the earth, in its vegetation, in its animal life, and that there is no way in which man has ever been able to escape the absorption of lead while living in this planet."[33]

While it is true that lead is natural, this fact is irrelevant to the question of whether it is toxic or whether human activities have led to widespread increases and exposure risks. True and irrelevant information is often heard in fractious arguments on environmental issues. The assertion that anything that occurs in nature cannot be harmful is also a frequent and erroneous refrain. Fortunately, it is fairly easy for the public to perceive this assertion as false today when it comes to heavy metals like lead, but the setting was different in the 1960s.

Patterson testified a week later. The scientific part of his testimony was measured, as it had been in his published scholarly work. It was also in direct conflict with Kehoe's statements. He noted that "the evidence for an increase in concentration in the blood of people in American cities is clear. . . . We can predict that the people in the cities will have higher concentrations of lead in their blood as a consequence of their absorbing the greater amounts of lead, and the difference [compared to rural areas] is due to the greater concentration of lead in the

air." When questioned about the health risks of low-level exposure, Patterson espoused a view diametrically opposed to Kehoe's: namely, that chronic poisoning was likely widespread in
America, "There is no abrupt change between a response and
no response. Classical poisoning is just one extreme of a whole
continuum of responses . . . to this toxic metal." Patterson's line
of reasoning is now known to be correct for many pollutants,
especially heavy metals such as mercury, cadmium, and lead, all
of which do occur in nature but are certainly toxic when humans
increase the amounts we take into our bodies.

Patterson's criticism of government reticence during his
testimony was biting: "It is not just a mistake for public health
agencies to cooperate and collaborate with industries in inves-
tigating and deciding whether public health is endangered; it is
a direct abrogation and violation of the duties and responsibilities of those public health organizations."[34] Congressional hearings often seem like much noise and no progress, but they wield
the soft power of public discussion, Senate pressure on federal
agencies, and exposure of risks. Muskie's hearing brought public
attention to the issue. More important, it put a great deal of pressure on the Public Health Service to pay more attention to lead.

Fledgling steps in developing a federal plan for some aspects
of air pollution were beginning under amendments to the ineffective Clean Air Act of 1963. Although these laws had few teeth,
they did at least begin the process of federal involvement in
municipal air pollution control programs, mainly through financial support for communities and for some research activities,
as well as the beginnings of some automotive exhaust controls.
Could lead in gas be included in such controls?

Muskie entered the following results of his hearings into the
Congressional Record to register his concern: "Lead in the atmosphere, attributable to automobile exhaust emissions, has

greatly increased. . . . It is well known that lead is a toxic substance, . . . not natural to the human body, and that, absorbed in large quantities, it can have toxic results. . . . We are concerned with the subclinical effects of long-term low levels of exposure." Muskie also regaled the Public Health Service to work much harder to investigate, and urged the broader US government to develop a better national organizational structure to deal with lead pollution in the air: "There is a need to . . . develop further information about subclinical effects of moderate lead concentrations."[35]

Science alone is never sufficient to solve an environmental problem, but it is always necessary. And it must be rigorously evaluated. In a time when there was no established mechanism for evaluation beyond whatever review of an individual study occurred prior to its scientific publication, the battle was fought to a draw between Kehoe's word and Patterson's, based on their own individual and deeply conflicting lines of evidence. Muskie must have read the following words into the *Congressional Record* with profound sadness: "There is no unanimity among the experts. . . . The subcommittee's hearings on environmental lead contamination raised many unanswered questions."[36] There was no more that he could do as a legislator, given this enormous disparity in scientific thinking.

Patterson left Muskie's hearing and went back to his work, making more measurements of lead in the ocean, in ice cores, and even in lead-soldered cans of tuna. He measured the lead content in canned tuna in his ultra-clean Cal Tech laboratory and found it to be ten thousand times that of fresh tuna. It appeared that the tuna casseroles eaten by millions of Americans were laced with hidden danger. He also managed to get a National Marine Fisheries Service (NMFS) laboratory to analyze a sample of the same wild tuna, and found that their analysis was so far off that they couldn't clearly tell the difference between canned and fresh tuna. The reason was simple: by this time lead was everywhere, including in the air in the fisheries' labs. When researchers dissected and handled the fresh tuna in their laboratory, their samples picked up lead contamination before they even tried to measure them. Patterson had always taken great pains to avoid contamination in his own ultra-clean laboratory, taking unusual steps such as the use of quartz rather than Teflon labware.[37] When Patterson's team dissected the fresh tuna in their clean labs and then gave it to NMFS, the results were profoundly different, showing much less lead. As Patterson described it, "You know Pigpen, in Charlie Brown's comic [strip], where stuff is coming out all over the place? That's what people look like with respect to lead. Everyone. The lead from your hair, when you walk into a. . . . laboratory. . . . will contaminate the whole damn laboratory. Just from your hair . . . from your clothing and everything else."[38] Ironically, the level of contamination was so high by the 1970s and 1980s that scientific measurements, even by unbiased researchers, were very likely to show high levels of lead that were utterly spurious.

The Environmental Protection Agency was established just a few years after Muskie's lead hearings, in 1970. Its origins have much to do with a growing environmental awareness in the American public in the 1960s, fueled in part by the publication of Rachel Carson's classic book, *Silent Spring*, with its unsettling account of the dangers of persistent pesticides. It was also inspired by increasing awareness of a trove of environmental hazards, including widespread air and water pollution.

By the 1970s, about two hundred thousand metric tons of lead were being emitted into America's air each year from automobiles and trucks.[39] But airborne lead was still not widely recognized as the grave environmental threat it was, and lead in the air was not among the topics that drove the establishment of the EPA. Nevertheless, the fledgling EPA's charge was to evaluate (and if necessary act upon) the fullest possible range of potential hazards. EPA had a duty to consider whether an approach could be found to reduce the amount of lead in gasoline.[40]

By 1972, research had shown that lead interferes with the body's production of hemoglobin and competes with calcium, ultimately affecting blood, tissues, and bone. Establishing the detailed biochemical mechanisms of damage are critical to establishing a clear path from pollution to health impacts. In this case, research had shown that lead literally inserts itself where other, benign substances like iron or calcium should be. It became clear that children were especially vulnerable to lead because their bodies are growing, calling into question the relevance of health studies using adult subjects.[41]

But in October of 1973, the Organization of Petroleum Exporting Countries (OPEC) announced the oil embargo as retribution against states that had supported Israel during its 1973 conflict with its Arab neighbors. The price of gas in the United States skyrocketed. Suddenly, getting gas involved waiting in long lines and paying high prices if you could get it at all. Changing anything having to do with oil or gas was now impossibly controversial.[42]

While phasing out lead in gas hit a roadblock in the early 1970s, beginning to deal with lead in paints gained traction around the mid-1960s, and continued to pick up steam in the years that followed. A push by children's health activists to screen blood for lead in Chicago schools revealed that up to 15% of children screened had dangerous levels of lead in their bodies. Shortly after this program began, two children were hospitalized with acute lead poisoning and died, amplifying news coverage and public interest.[43] Lead was finally becoming personal, although some found it easy to continue to blame the victims, since it seemed that so many of them were people of color.

Screening programs began in several other cities, promoted by activists like the Young Lords in New York. In a number of large American cities, 25%–45% of children between one and six years old living in high-risk areas had excessive levels of lead in their blood.[44] Peeling lead paint was thought to be the main source of childhood lead poisoning in the 1960s, particularly in badly maintained homes in poor neighborhoods. Paradoxically, even Kehoe acknowledged lead in paint as a threat to children: "The occurrence of lead-containing commodities and the use of lead paints on furniture, toys, and other objects within the reach of small children is much too common to ignore."[45]

While there was growing recognition of the danger, states and cities did not manage to take the primary prevention step of removing lead paints from homes. Some places designed policies, but nearly all were quickly abandoned because landlords and homeowners resisted them or they proved ineffective. Massachusetts managed only a meager 0.5% decrease of its leaded paint homes from 1982–1986, while Philadelphia established an ordinance but retracted it.[46] Landlords argued that if they took on such costs, they would have to raise rents, which would make housing less affordable for the nation's poor. And the balance between such costs and the massive benefits to public health that would have accrued were not part of the thought process. Instead, cities, states, and the federal government turned to broadening the screening of children for excessive blood lead levels, in an effort to at least monitor and possibly head off the worst cases.

At that time, many poor neighborhoods with peeling paint were black communities, so the innocent children of the black population who could not afford better homes were disproportionately subject to lead exposure. Lead paint became an issue that struck at the heart of discrimination and inequality. When the Vietnam War, the environment, women's rights, and civil rights utterly transformed the national conversation beginning in the 1960s, the prevalence of lead poisoning in minority children catapulted the issue to attention as part of a rapidly growing social and political upheaval.[47]

With so much public attention on the problem, the US government finally responded to the issue in the late 1960s by greatly expanding the screening programs for children through community health services and funding from Medicaid. Within a few years of these measurements, it became clear that a frightening number of children were at risk from lead damage. Public attention to the issue inspired congressional hearings and the passage of the Lead Paint Poisoning Prevention Act of 1970, which again skirted the delicate issue of de-leading existing structures but resulted in increased federal support for screening children's blood, funding very broad surveys of millions of children in the 1970s. The surveys revealed that lead poisoning was rampant in the United States, not just in poor or black neighborhoods, and not just in urban areas but in smaller cities and even some rural areas. No longer was this just a civil rights issue. Finally, lead became a personal and perceptible issue for each and every American parent.

The executive branch of the federal government was once again slow to act, in part because of its agency structure and organization. It simply did not have the authority to outlaw lead in paint—only the legislative branch could do that—but it did have the prerogative to stop its future use in federally funded structures, such as public housing for the poor. It passed a law eliminating the use of interior leaded paints in "residential structures constructed or rehabilitated by Federal Government or with Federal assistance" in 1971. While such a step may seem modest, it also put pressure on manufacturers by reducing the market for lead paint and fostering development of other options. Since the government is a large customer, this is a frequent tactic to promote change. For example, one perhaps small but very simple way the US government is promoting the transition to electric vehicles is simply to do it for federally-owned vehicles. The newly formed Consumer Products Safety Commission banned lead in painted toys and furniture shortly thereafter. The Commission, like the EPA and other new agencies, had a lot to learn before it could do more. Occasional violations continue to occur: for example, one company was caught selling products coated with 60% lead in 2009, but the Commission has the power to enforce fines.[48]

Another environmental issue gained the necessary traction to be at the forefront of national thinking in the late 1960s and 1970s, and that was the thick air pollution blanketing many of America's cities, most notably Los Angeles. This led to the adoption of the catalytic converter and with it the phaseout of leaded gas. At the same time, concern about lead in paint prompted widespread measurements of children's blood lead levels, and those declined in an unexpected way as leaded gasoline use was reduced, irrefutably demonstrating the connection between leaded gasoline and children's health and thereby accelerating the demand to remove leaded gas. In a remarkable and circuitous way, smog and lead in paint therefore ultimately saved millions of the nation's children from lead poisoning from gasoline.

Congressional legislation doesn't often specify dates and numbers, usually leaving that to the implementing agencies to work out. But the Clean Air Act of 1970 is a remarkable exception. Among other things, the Act mandated a 90% reduction in emissions from light-duty vehicles (cars and trucks) of smog-forming carbon monoxide and nitrogen oxides by 1975 and 1976, respectively. Putting such specific language into federal air pollution legislation was essentially unprecedented, but Muskie would not budge on it during negotiations.

While its very aggressive timetable had to be relaxed a bit in later years, the Act sent a stern message to auto manufacturers that business had to change, and change fast. The only practical pathway to such a large reduction in emissions was the catalytic converter. The necessary catalytic converters are, however, fouled by lead. Lead would have to be removed from gasoline if new cars with catalytic converters were to meet the new smog control requirements. Muskie pushed the Act through to force the auto industry to comply quickly, before political will weakened or alternative strategies on smog control could be advanced (such as a greater focus on power plants and a reduced one on automobiles). The Clean Air Act stands today as one of the clearest examples of how technology-forcing policy can mandate the development of the technology and innovation that is sometimes needed to transform what may seem impossible into what is practical.

EPA staffers went back to the drawing board on lead poisoning while the drama of the Clean Air Act's specifics unfolded. They wisely pivoted to pick up on the growing concerns about children's health with regard to paint and blood lead. Growing evidence aided the effort, especially findings from several studies that lead was contaminating soils as well as household dust in many parts of the nation. Children naturally play in dirt and dust, and were clearly at risk. The EPA argued for reduced lead in gas on this basis, requiring a lower amount of lead in the gas that would be sold for old cars alongside the unleaded gas now necessary for the new cars. Industry attacked their lead standard, but it survived legal challenges, because the underlying science came

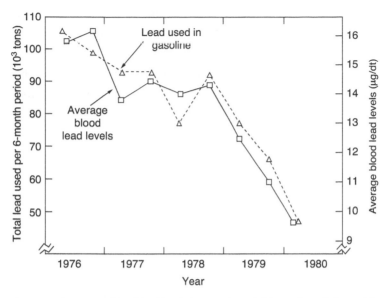

Decreases in average blood lead levels in American children, plotted along with total lead used in gasoline for each six-month period, from 1976 to 1980.

from multiple studies, and because the public outcry about lead risks to children's health was strong enough.[49]

As the technological and political dimensions of the nation's cleanup of America's air unfolded, a remarkable and unshakable truth emerged. As the nation reduced lead in gas, researchers continued to analyze those screening samples nationwide to look for dangers from paint. The samples did identify some children subject to increased risks from the paint in their homes. Shockingly, they also revealed lockstep lowering of the average blood lead in the nation's children as the sales of leaded gasoline declined. It became crystal clear that most of the lead in children's blood across the nation came not from the paint that had inspired the blood lead screening in the first place, but from the fog of auto exhaust in air inhaled all across the country.[50]

The phaseouts of lead in both gas and new interior paint followed. The Consumer Products Safety Commission decreased and eventually banned the use of interior leaded paints nationwide, beginning in 1978,[51] and the EPA steadily lowered the standards for allowed lead in gas. Turnover of the fleet followed a stylish national zest for buying new cars frequently, with each car equipped with a shiny new catalytic converter and the requirement of lead-free gas to go with it.

Research showed that not only the soil but also the blood of children living close to major highways was higher than those found farther away, implicating leaded gas as a widespread threat to children, albeit one difficult to perceive because many children were asymptomatic. As early as the 1940s, some researchers had suggested that unrecognized lead poisoning affected school failures and behavioral problems in children, but findings were debated.[52] As epidemiological studies advanced through important strides in their methods and scope, researchers could finally detect more clearly just how damaged children were throughout the nation, not just in poor minority communities.[53] For example, researchers could now quantify the impacts of lead on children's development and performance in school while accounting for the previously sticky factors that could confound such studies, such as nutrition, parental education levels, socioeconomic status, and more. With the advent of computers and modernized data-analysis methods, epidemiology and biostatistics had developed the tools to comprehensively attack the issue in ways that clinicians in the early twentieth century could not have imagined. In 1979, Herbert Needleman published a study of lead impacts on children that advanced the field. He collected freshly shed baby teeth rather than blood samples, in order to obtain a measure of their cumulative rather than just current exposure to lead. The study involved mainly white children from the suburbs of Boston. Carefully controlling for confounding factors, he found a significant relationship between high lead levels and poor performance in school. He also documented an apparent impact on IQ, as well as physical reaction times. In short, high lead levels produced impaired children.[54]

We will never know how many children suffered and how exactly lead affected a generation of Americans, especially poorer Americans. While scientific debate continues regarding whether or not even small amounts of blood lead can reduce IQ, the range of health effects documented in studies by Needleman and others prompted a rapid lowering of the blood lead levels deemed safe by the Centers for Disease Control and supported a more rapid phaseout of leaded gas.

By the late 1970s, public outcry about health effects along with new inventions meant that tetraethyl lead gas additives in cars were phased out. Computer control and electronic ignition made the transition to catalytic converters and lead-free gas easier for new cars, since with these advances, engine knock could be controlled by automatic adjustments in ignition timing instead of using lead additives. By 1980, policies had ramped down gasoline lead levels (average of leaded and unleaded sold) in the United States by more than 50%.[55] By 1996, leaded gasoline was no longer allowed in the United States, and children's average blood lead levels had fallen by 80% compared to 1976 levels.[56] The United States was a world leader in attacking this problem, and an analysis of the reduction in blood lead levels in various countries demonstrates that the speed of the US process in this environmental action met or exceeded that in almost every other nation.[57]

The frameworks to clean up America's air and water were part of a more ambitious revamping of the federal government that included the creation of the Environmental Protection Agency (which helped advance the phaseout of lead in gas) and the Consumer Products Safety Commission (which finally banned lead in interior paint). Without the right governmental structure, it's just hard to get things done. Remarkably, a few countries continued to use leaded gasoline well into the twenty-first century. Some of the holdout countries were run by dictators, such as North Korea and Myanmar, where many of the governmental structures needed to safeguard the environment and limit associated risks to public health are essentially non-existent. Nevertheless, even these two eventually stopped selling leaded gas. Some leaded gasoline was still sold as of 2018 in Algeria, Yemen, and Iraq as well. It is not easy to discern exactly why, but war, conflict, poverty, and oppression of public discourse are among the things that surely make environment and public health a distant topic on any national agenda. But in a triumph for improving public health, the UN announced a worldwide end to leaded gasoline sales in 2021.[58]

Even though the use of leaded gas and paint ceased entirely in the United States, the problems they created are not entirely behind us. Dangerous lead legacies linger among us: in many tons of contaminated soils (particularly near highways, storage facilities, and chemical plants) and on the walls of many older American homes. As recently as 2000, twenty-four million US homes still had significant indoor lead paint hazards, especially in the Northeast, in homes built before about 1970. It is now clear that Hamilton and Gibson were right to be concerned with lead dust. The lead that went into engines for decades formed dust that wafted so widely that it could be seen in Greenland's ice. Apart from peeling paint, the key source of lead, even in well-maintained homes, is the dust generated by friction in window and door jambs. Replacing lead-painted windows and doors is one way to reduce this hazard.[59]

Among all environmental challenges, we have the hardest time undoing expensive investments in assets that we've already installed and plan to use for a long time. It's fairly easy and cheap to decide to stop selling lead paint, but much more costly to remove it from the millions of walls, windows, and doors that it coats.

Even in the twenty-first century, children still suffer from lead poisoning, not only from remaining lead paint and tainted soil but also from lead water pipes, a problem that is far more costly to fix. But price doesn't always stop progress when public demand is strong enough. Once again, many of the worst incidents of lead in pipes and municipal water have occurred in urban areas populated by minorities (notably Flint, Michigan, but also others). Flint is a majority black city where 40% of its families lived below the poverty line. When the city fell into bankruptcy in 2014, an unelected, state-appointed emergency city manager made the ill-fated decision to switch the municipal water supply to an outdated one that proved to be unsafe. Citizens' repeated complaints were ignored for eighteen months—a public health outrage that would be highly unlikely in richer communities.[60] As in the civil rights movement of the twentieth century, the focus on inequality has been boosted by concerns around environmental justice and the public activism of the twenty-first century Black Lives Matter movement. In 2021, Congress passed a long-awaited American infrastructure bill that included provisions to remove all of the remaining lead water pipes nationwide.[61] And although the bill was controversial in some ways, bipartisan support propelled the lead pipe removal measure to near the top of the list throughout the debate.[62]

The phaseouts of lead in gasoline and of lead in paint are rightly hailed as two signature triumphs of environmental action, both in the United States and internationally. A few individuals were pivotal: those who advanced the science essential for public understanding, including Gibson, Hamilton, Needleman, and especially Patterson, as well as Senator Muskie, who took clever advantage of widespread public support for environmental action to spearhead the skillful crafting of the Clean Air Act of 1970. Revelations about the lead in children's blood made the problem personal rather than abstract. After thousands of years of confusion, the profoundly personal health dangers of lead became perceptible to people as the science advanced and became easy to understand. Practical substitutes swiftly began to be available because of policy decisions. These ingredients ultimately framed the twin trajectories of lead in paint and in gas during the twentieth century.

You may wonder if Ed Muskie was a genius who slew both smog and leaded gas, two of America's most dangerous pollutant dragons, with one clever stone, or just lucky. I'm inclined to go with the genius theory. Another, much more reluctant and shy genius paved the way for that success through her key role in awakening America to the very concept of silent environmental risk a few years before, when pesticides began perceptibly ravaging America's birds.

5

Pesticides

TAMING WEAPONS OF WAR

DUXBURY, MASSACHUSETTS

The woman's property included acreage at the edge of a marsh. Watching the local birds was her passion, and she rejoiced in the numbers and variety attracted to her feeder. But the day after the mosquito-control plane crisscrossed her land spewing a fog of pesticide, she found not one but seven dead and dying birds.

She angrily penned a letter to the editor of the *Boston Herald*: "We picked up three dead bodies the next morning, . . . (and) the next day three were scattered around the bird bath. They all died horribly; . . . bills were gaping open, and their splayed claws were drawn up to their breasts in agony." When that was done, she wrote to a friend, describing her "blistering letter."[1]

The year was 1958, and the friend was Rachel Carson. Carson later wrote that receiving the letter "brought my attention sharply back to a problem with which I had long been concerned." That letter, she said, was what made her write a book that would ultimately become a turning point in the relationship of humankind to the environment.[2]

As Americans heaved a collective sigh of relief at the end of World War II, they were awestruck by two innovations that paved the way to the surrender of Japan. They credited the powerful and horrifying new weapon, the atomic bomb, with shortening the war, albeit at the cost of horrific death and devastation on the Japanese side. The development of the bomb would forever endow humanity with the tools of our own terrible destruction. The second innovation was a remarkable and efficient weapon against the insect world: a chemical called DDT. Although we didn't know it yet, this one also contained the seeds of global catastrophe.

Some wars have been won or lost because of better catapults, more powerful cannons, and bigger bombs. But scholars argue that more conflicts have turned on the impacts of insect-borne diseases. Many an invading force has found its most dangerous enemy to be their own lack of resistance to local diseases spread by pests, along with the overcrowding and lack of sanitation often associated with large marching armies.[3] Typhus is distinct from the better-known typhoid fever, and is spread by insects—particularly lice, chiggers, and fleas. Modern microbial analysis of ancient remains shows that Napoleon's Grand Army suffered from epidemic typhus as they retreated from their failed assault on Russia.[4]

The threats of typhus and mosquito-borne malaria loomed large for America's forces in World War II until military scientists widened their chemical toolkit. DDT was to earlier methods as the machine gun was to the flintlock. The chemical created a firestorm in the nervous systems of nearly every bug, resulting in spasms, erratic motion, and death. It could be easily distributed in sprays, powders, or even dispersed in oil, and the persistence of its killing power was unmatched, lasting not just hours or even days but months. In short, it was the A-bomb of insecticides.

The US military issued repellents containing DDT to soldiers to safeguard them during night combat, and spraying via airplanes eradicated mosquitoes over entire islands even before ground forces arrived. DDT's potential for peacetime benefits for agriculture and human comfort (as well as disease) could only be imagined.

Among DDT's postwar insect targets was the gypsy moth, whose feared swarms could defoliate forests and apple orchards. By 1947, it was clear that aerial spraying of DDT could keep gypsy moth populations down.[5] Entomologists began to look to other battlegrounds, such as the cattle tick and screwworms that plagued livestock, the boll weevil that damaged the cotton crop, the argentine ant that threatened citrus fruit, and the housefly and horn fly. Dairymen who sprayed DDT reported no flies in their barns for up to six weeks. A "No Flies in Idaho" program promoted by university scientists took DDT spraying up a notch to the statewide level. Idahoans so enthusiastically embraced the campaign that it became difficult to trap even a single housefly in the state.[6]

The possibilities seemed limitless, although some scientists did caution that beneficial species might also be threatened, prompting testing. The standard regimen for impacts on animals and plants in the wild involved inventorying beneficial species in a given area (such as a forest), spraying once, and inventorying again. If changes seemed minimal, the test was deemed a success, and the substance safe.[7] It seems remarkable to look back on this with the benefit of the knowledge gained over the decades since and note that whether repeated applications might be more dangerous than just one, and whether delayed responses could occur were questions left entirely unaddressed.

American farms became larger and more industrial after World War II, fueling widespread increases in the use of DDT. Farmers snapped up surplus war planes, so they could crop-dust generously—usually with the new miracle insecticide. Public health programs also embraced the new chemical. Even before the war was over, clouds of DDT appeared over the malarial regions of the American South. At first the targets were restricted to military bases, but within two years the program covered all areas where malaria was found, over a dozen southern states. It was not long before some private citizens living near sprayed areas began to complain of falling ill.[8]

The scholarly literature had already documented that humans could indeed die or be sickened after DDT exposure. One study published in the respected *British Medical Journal* in 1945 described a dramatic experiment in which the author himself and his technician exposed themselves to large amounts of DDT, much more than likely would be experienced by consumers or even by farmers. The two men sat in a chamber coated with DDT for two intervals of forty-eight hours, separated by forty-eight hours outside the metal vessel. They monitored their own health for an additional 47 days and scrutinized their condition before and after exposure, including blood, urine, and neurological tests. Both experienced extreme tiredness, sore throat, irritability, vision problems, tremors, and joint pains so severe that one went for an X-ray. Symptoms persisted for twelve days after exposure, but gradually subsided to zero over about the next month. While both men recovered, the paper concluded by emphasizing that "DDT intoxication in humans is a hazard to be considered and guarded against," while the acknowledgments thanked the technician for his "cheerful collaboration as an experimental animal."[9] The article that followed covered a Kenyan child's convulsions, tremors, and death within hours after accidentally drinking DDT. The latter was not the first account of a death from DDT, although it was one of the saddest.

The British study was among many to prove that there was little danger of acute toxicity from an exposure or even two to DDT for a limited time—if used according to stipulated directions. The circumstances that led to immediate illness and death in these laboratory situations or the case of the Kenyan child were deemed vastly different from those of "proper use." Industry used these results to emphasize that DDT just needed to be used "responsibly," characteristically casting the blame on the user rather than the substance if any problems arose. Just as in the case of leaded gasoline additives, victim-blaming was an industrial strategy that helped keep the risks from becoming perceptible for the public.

This research, while conducted according to the standards of its day, again left holes in knowledge that are astounding by later standards. One can only guess what health effects people living near farmlands might have experienced following fogs of DDT wafting to their homes not just once, but repeatedly over several years. The available data simply did not address the question of whether DDT could accumulate in the bodies of people or animals, and whether chronic conditions in humans might be caused by the cumulative effects of repeated doses (albeit likely of smaller amounts) over years or decades.

DDT and other wartime chemicals newly available to the public received some negative press, and a few states, including

Missouri, California, and New York, issued warnings that these substances, like other pesticides, were poisons.[10] But home gardeners were well used to such warnings, and national sales of DDT continued to soar.

Why wasn't the federal government stopping or at least slowing the potentially dangerous use of DDT? The EPA had not yet been born. Two federal agencies were jointly tasked with monitoring and regulating the use of insecticides and pesticides in agriculture: the Department of Agriculture (USDA) and the Food and Drug Administration (FDA).

In June of 1949, scientist and commissioner of the FDA Paul Dunbar gave a speech about his agency's work to the National Agricultural Chemical Association, the industry group that included the manufacturers of pesticides. Dunbar told them that "by and large insecticides are poisons. If they were not poisonous they would be of no value." By this time, a raft of studies had documented deaths and illness in laboratory animals exposed to DDT and its chemical relatives, including deaths in guinea pigs and rabbits, as well as death and liver damage in rats, and Dunbar took care to note such work. He emphasized that under federal law, poisonous or dangerous additives to food were to be banned.

But Dunbar was acutely aware that the law of the day included a special set of caveats on banning insecticides whose elimination would "severely curtail food production." The FDA's main role in that special case was to set safe tolerances for how much residue could be allowed in foods. It was a mammoth job. While DDT was the most widely used new chemical, it was accompanied by half a dozen sister organochlorine molecules with similar structure and toxicity to insects or fungi. These included dieldrin, heptachlor, and chlordane.[11]

In some cases, researchers didn't even have methods to measure the residues on food. A great advance in chemical science to do that was coming (the vapor phase chromatograph) but it was about fifteen years away.[12] So much was unknown, including how easily any surface residues could be removed by washing, whether the chemicals were absorbed into the plants, and more. So although the FDA's responsibility was to consider not just acute poisoning but also chronic poisoning and cumulative toxicity, there was simply not enough scientific information to do that. And if there were no official restrictions, then the door was open to a wild-west approach with no limits at all. In short, the law was set up so that agrochemicals were assumed innocent of harm until proven guilty, and there was nothing the FDA could do to stop it.

Dunbar noted "increasing danger of exposure of the general public to small continued intakes of DDT" and emphasized that "we do not know how serious this hazard is in terms of human damage." He exhorted his audience to exercise "discretion and discrimination." Under the law, urging the industry to voluntarily reduce or limit use was all that he could do. The soft power that he wielded of exposing potential dangers was largely ignored by industry, with one important and deeply personal exception for parents: dairy.

A few months before the speech, headlines across the United States had carried the unsettling news that DDT could be found in the milk drunk everywhere by the nation's children. The FDA was asked to evaluate if DDT could be used safely in the dairy industry. Here Dunbar drew a line in the sand: "The FDA cannot and will not set up a tolerance for DDT in milk."[13] Public uproar raged, and the industry backed down. Flies returned to dairy barns in Idaho and elsewhere in short order.

History shows that new laws, new agencies, and changes in the authority of federal agencies were coming, circumstances that would profoundly alter the world for agricultural chemical companies. But as his audience listened to Dunbar, there is little doubt that some privately thought that there was no need for them to be troubled at all: they could certainly count on a different federal agency—the USDA—to keep them insulated from any more need beyond that of the dairy industry to dwell on niceties like "moral obligations" and limit use, as the FDA recommended.

In 1947, the US Congress passed the Federal Insecticide, Fungicide, and Rodenticide Act, which vested the USDA with the legal authority to regulate pesticides. The USDA had long been charged mainly with promoting the profits of US agriculture. Relationships between the USDA's leaders and the farmers, as well as with the agricultural chemical industry, were warm and friendly to say the least. The USDA's jobs under the Act included the legal registration of pesticides, a required step before a substance could go on the market. But the registration existed to protect the farmers' pocketbooks and not the public. The USDA tested to ensure only that any registered pesticides killed the bugs as advertised. There was no requirement to evaluate human health or ecosystem impacts.[14] If a pesticide or fungicide was thought to be dangerous to the public, it could still be "registered under protest," and distributed at will.

With the FDA largely hamstrung in its authority, except in rare cases when tolerances could be scientifically established, and the USDA leadership acting essentially as a willing arm of the farm and agricultural chemical lobbies, the American consumer was exposed to a growing load of lightly tested agricultural chemicals for decades. DDT was just too good at killing bugs. Its practical uses were quite diverse, not only in farming but also in private homes and in the wild lands that were breeding grounds for pests. The precautions that might have kept its use to a minimal level with safe handling fell by the wayside.

But rumblings of concern about DDT's safety for wildlife and people nevertheless began to be reported in scientific studies, showing that it is awfully hard to keep good science down. In 1946, US Fish and Wildlife Service scientists Clarence Cottam and Elmer Higgins published a review in the *Journal of Economic Entomology* reviewing available studies by many scientists. The individual papers recounted many different and disturbing losses of wildlife after the use of DDT, including wild birds killed during gypsy moth spraying, loss of fish following use of DDT to control insects along a stream, and death of blue crab after spraying for mosquitoes in brackish water in coastal New Jersey.[15] And technical reports of DDT's risks (for example, dangerous accumulation in soils and risks to beneficial insects) continue to pour in from scientists in many different agencies (including USDA's own scientists, despite the close association of the agency's administration with farmers).

But the distribution of these findings, and nearly all others of their day, was limited to the community of scientists engaged in such studies and seldom reached the public eye. These were the days before scientific results were distributed to the public via social media or university and government agency press releases. Scientific knowledge spread via densely written research papers published in obscure journals. Most members of the public in the 1940s learned about scientific developments from daily newspapers or books. In 1945, the *New York Times* reported that Lucille Stickel, a junior biologist with the US Fish and Wildlife Service, had found that DDT had an extremely toxic effect on fish and other cold-blooded aquatic creatures when used against mosquitoes in stagnant waters (such as lakes). But this article only appeared because the *Times* chose to report on it in a novelty piece titled "Women Scientists Advance Wildlife Studies." Stickel's work was listed along with that of other women who studied such things as stretching the meat supply using wild game, and how to raise animals to produce the best fur.[16] The far more frequent focus of news stories about DDT and related chemicals prior to 1962 was some new and wondrous use showing how marvelous these new pesticides were, such as clearing boats of barnacles.[17]

DDT became the tool of choice in the battle to save America's beautiful shade trees from a disease discovered in the Netherlands and first found in America in 1930, dubbed Dutch elm disease. After yellowing leaves appear, death of the elm tree often follows. Beetles carry the deadly fungus from one tree to another, so killing the bugs effectively can stop the spread of the disease. Emulsion spraying of a solution of 2% DDT seemed to be the ideal countermeasure to check the blight, and American communities quickly adopted it, particularly in the Midwest, where the disease was running rampant by the 1940s.[18]

"Tremoring robin (female) . . . brought to office by students . . . died a few hours later." Professor George J. Wallace of Michigan State College (later University) documented an epidemic of robin deaths over a four-year period in a paper in *Audubon Magazine*. The campus population of robins dropped from 185 birds in 1954 to none or a few by 1958. He did another census in 1959, from which he concluded that "obviously the campus is serving as a graveyard for most of the robins that attempt to take up residence in spring. . . . To date about 45 different communities (all with Dutch elm disease control programs) in the Midwest have reported a similar dieoff."[19]

About three hundred miles away, on the campus of the University of Illinois in the twin cities of Champaign and Urbana,

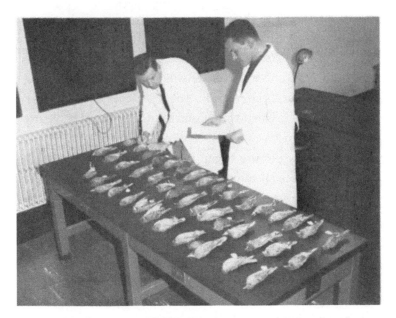

Two scientists examine dead robins picked up on a university campus in 1959 (from *Audubon Magazine*, Spring, 1960). The birds are still in the museum on the campus of Michigan State University.

students and scientists also found stunning numbers of robin corpses. Few robin deaths had been observed in the summer of 1949, after the first season of spraying the trees for Dutch elm disease. But the shock came the next spring, when "the numbers of dying robins . . . attracted unsolicited attention. Dying birds were reported frequently after rains. . . . Most dying robins exhibited typical tremors associated with DDT poisoning." How could this be so when the 1950 spraying season had not even started?[20]

The Illinois scientists measured not just DDT but also the derivative molecules that it can form in the environment and in the bodies of animals and humans. At first, they suspected that the pesticide must be contaminating the puddled water that the robins drank, but analysis of the water showed low concentrations. Then they measured DDT and its derivatives in the brains, tissues, crop and gizzards of the birds. These revealed startling amounts, especially in the digestive systems. Any observer of robins in a wet climate will have noticed their enjoyment of earthworms after a rain. The researchers also measured these chemicals in earthworms, and they determined that earthworms could concentrate an incredible amount of DDT.[21]

Suddenly, the picture began to emerge: rain had rinsed the DDT-soaked trees all summer long in 1949, washing the chemical down to the ground. Remaining DDT on the leaves had also fallen in autumn to form leaf mulch during the autumn and winter. The food for the earthworms in turn during the next spring was the contaminated soil, mulch, and leaf litter. In short, the robin's favorite food was brimming with poisons from the spraying of the past year, even before another deadly load was added.

But why didn't the DDT decompose as summer turned to autumn, and autumn to winter and finally spring again? This simple study pointed to the long lifetime of DDT in the environment—a characteristic that amplified its deadly implications. Quantitative work would ultimately show that DDT could persist in soils not just for a season or a year after application, but for twenty years and more. While nowhere near as long-lived as lead in soil, DDT wasn't just an effective pesticide—it was also a very persistent pesticide. And persistence is a terribly dangerous property in any risky chemical released to Earth's environment. It means that the loading of the hazardous substance will build up systematically if it is applied in equal amounts year upon year, with the second year's amount nearly double the first, and the third nearly triple and so on. High rates of compounding interest have destroyed the lives of many a poor borrower, and the same vicious principle applies here.

If a chemical with a twenty-year lifetime is applied in equal amounts at the same time once per year, then at the end of ten years the environmental load is almost eight times that of the first year's application. And if the use of such a chemical were abruptly stopped after ten years, it would take about forty years for the environmental load to return to what it was after the first year. In contrast, if a pesticide with a one-year lifetime in the environment is applied every year in the spring, then much of what is applied decays away between applications, and in a few years the environmental load that remains before the next year's load is applied reaches a steady value of about 60% of the peak load. In the first case, natural systems may well collapse under the growing load, while in the second, they are far more likely to be unaffected or adapt. Despite its simplicity, this fundamental principle was not understood in the early years of studies of DDT and its sister poisons.

On top of persistence, DDT has another dangerous property that further amplifies its effects on wildlife, particularly predators. It dissolves and concentrates in fat. The fatty tissues of the earthworms stored up the DDT that fell from the campus trees to the ground. So as the earthworms munched their way through wet leaves and soils impregnated with larger and larger loads of DDT residues from previous years, their bodies accumulated more and more DDT. And the loading became even more concentrated in robins, because the birds voraciously eat earthworm after earthworm, accumulating more and more DDT in their own tissues. Baby robins eat about half their own body weight each day. Consuming the DDT-laced insects rapidly built up the DDT in their bodies to toxic levels.

Wild inhabitants of ecosystems worldwide reeled under the growing weight of DDT amassing around them in the years after

World War II. Persistence meant that even careful use of low concentrations of DDT to kill specific bugs was an ecological time bomb. And careful use was a rare event. Much more than necessary was usually applied, because DDT was cheap, and ensuring the killing of every last bug was tempting. Researchers only recognized these risks over time, as scientists became skillful enough to perceive DDT's trail of death and damage and perhaps most importantly, as that science emerged from obscurity into the public sphere, prompting citizens' own observations and heightening concern. That brings us back to a blistering letter about very perceptible bird deaths sent to an important friend.

Rachel Carson was born on a farm in Pennsylvania in 1907. Her exceptional writing abilities emerged early. She wrote her first published story for a children's magazine in 1917. In an era when many occupations and colleges were open only to men, her mother Maria encouraged the young Rachel in literary pursuits, envisioning a socially acceptable career and livelihood for her gifted daughter as a writer. Maria fostered Rachel's goals not only with motherly love and friendship, but also by giving piano lessons and selling the family silver, which allowed Rachel to attend college in an era when that level of education was largely attained only in far more prosperous families. Rachel began her studies as an English major at the Pennsylvania College for Women in Pittsburgh, with a tuition scholarship she earned through an academic competition. The Carsons still had to sell some of their farmland to help with room and board fees, and Rachel accumulated some college debt that she would need to repay.[22]

At college, Rachel realized that her love for science was even stronger than her desire to be a writer, and she switched her major to biology. She set her sights on an academic job, despite the fact that successful careers in research or teaching were virtually closed to US women in the 1920s. By December of 1928, Rachel was applying for graduate school at Johns Hopkins, where she would ultimately add a zoology master's degree to her bachelor's in biology. She received admission along with a full graduate tuition scholarship ($200 for one year).[23]

Before entering Hopkins, Carson obtained a summer post at the prestigious and intellectually stimulating Marine Biological Laboratory at Woods Hole, Massachusetts, on the shores of Cape Cod, where she discovered endless ways to expand her growing love of the sea. She was fond of walking the shore to examine the myriad inhabitants of tide pools and watching polychete worms mate under the dock by the light of the full moon. She "began storing away facts about the sea."

Carson's MS degree was awarded in June of 1932, just as the Great Depression deepened. She intended to go on for a doctorate but was forced to drop out of school to help support her family in 1934. Shortly thereafter, her ailing father died, and Carson became the sole support for her mother, sister Marian, and Marian's two children. Desperate for a steady job, Carson turned to a friend from college who was working as a government scientist. Through this contact, she was introduced to Elmer Higgins of the Bureau of Fisheries, who needed someone to help write fifty-two short radio pieces on marine life called "Romance Under the Waters" for $6.50 per day.

Carson's radio pieces were a lively and engaging success, and she went on to a variety of writing tasks for payment with the bureau while hoping that a federal position would open up. In July of 1936, she finally received a federal appointment as a junior aquatic biologist with the Bureau of Fisheries, at a weekly salary of $38.48. Her work included scientific analysis of statistical data on fish populations, along with writing reports and producing public brochures on conservation. Carson's mother, sister, and two nieces moved with her to her new job in Silver Spring, Maryland. In 1937 her sister died, leaving her as the sole support for her aging mother and two nieces.

Her mentor, Elmer Higgins, suggested that her article for the Bureau of Fisheries about the undersea world was so remarkable that it ought to be published somewhere other than a government brochure. A few sentences from the essay readily demonstrate why: "To sense this world of waters known to the creatures of the sea we must shed our human perceptions of length and breadth and time and place, and enter vicariously into a universe of all-pervading water. For to the sea's children nothing is so important as the fluidity of their world. It is water that they breathe; water that brings them food; water through which they see, . . . water through which they sense vibrations equivalent to sound." Carson was successful in submitting an expanded version of the work as an essay to the *Atlantic Monthly*. Her career as a nature writer was launched when the essay attracted the attention of a book editor at Simon and Schuster.[24]

In 1940 the Bureau of Fisheries became part of the US Fish and Wildlife Service, and Carson continued to write government brochures along with articles in popular magazines. Her 1941 book, *Under the Sea Wind*, was excerpted in the *New Yorker* in advance of publication. This first work by the "poet of the sea" received broad critical acclaim and praise for its scientific accuracy but had only limited sales when the Japanese attack on Pearl Harbor struck just a month after its publication. When Carson's second book, *The Sea Around Us*, appeared, it became a bestseller and an immediate success, winning the 1952 National Book Award for Nonfiction. Her first book was republished as a result of this new-found fame, and it also made the best seller list at the same time. The *New York Times* described this as a "publishing phenomenon as rare as a total solar eclipse. . . . Great poets from Homer . . . to Masefield . . . have tried to evoke the deep mystery and endless fascinations of the ocean; but the slender gentle lady who is editor of the United States Fish and Wildlife Service seems to have the best of it. Once or twice in a generation does the world get a physical scientist with literary genius."[25]

In an era when television had not yet penetrated the American living room, the power of the book made national reputations, and Carson became the foremost naturalist of her time: 1952's version of Sir David Attenborough's multi-decadal stardom via his documentary films of the late twentieth and early twenty-first centuries. Carson became wealthy enough to quit her job at the Fish and Wildlife Service and devote her time entirely to writing and family. She had often vacationed with her mother, sister, and nieces at rented cottages on the Maine coast. With her new-found wealth, she determined to own a piece of that magical shoreline. She purchased land along the Sheepscot River near Boothbay Harbor and built a cottage.

While at the Fish and Wildlife Service, Carson saw the disturbing reports by colleagues such as Lucille Stickel about the strange losses of wildlife linked to the use of DDT and other pesticides. In her role as information specialist, Carson was in frequent contact with Fish and Wildlife Service scientists studying the impact of pesticides on wildlife, and had many conversations with her supervisor and mentor, Elmer Higgins, who had been among the first to summarize the science of the issue in 1946.

As one of the nation's foremost nature writers with a widening range of contacts not only from her many years at the Fish and Wildlife Service but also with the National Wildlife Federation, news media, and readers of her books with whom she corresponded, Carson followed with interest a wide range of environmental issues. Members of the Audubon Society wrote to her to communicate growing uneasiness about USDA's plans to spray pesticides over thousands of acres in order to combat the fire ant in 1957. She also heard from a group of Long Island citizens who brought suit to halt the spraying of DDT on their properties for mosquitoes and Dutch elm disease. The Long Islanders were well organized, and the trial was replete with testimony and scientific evidence regarding potential damage not only to birds, bees, and fish, but possibly the health of children.[26]

Among other evidence, the Long Island trial led Carson to the story of the robin deaths documented by Wallace at Michigan State College. A larger picture was slowly building, as study after study revealed hazards not only to birds, beneficial insects, and fish, but even the occasional deaths of a few unlucky consumers or farm workers. She decided to write something about the dangers, perhaps an article or book with a coauthor, or possibly an edited book with several coauthors.[27]

Despite the wide exposure of the Long Island trial, something was missing before science could form a basis for environmental decisions. We did not yet have today's vital tool for setting environmental policy, known as scientific assessment, in which a broadly diverse team of qualified scientific experts and reviewers evaluate all the published scientific work on a subject. Rigorous technical training of the experts underpins successful assessment today, along with review by many additional qualified scientists in an open process. Typically, the experts and reviewers include some scientists who are involved in the studies and others less so or not at all, providing for both deep knowledge and unrestrained criticism. Modern scientific assessment provides a synthesis of science and policy bottom lines by identifying what is known and what is not in language understandable by, and disseminated to, the public as well as to policymakers. Scientific assessment would later become a cornerstone that led to the

Montreal Protocol to protect the ozone layer, and a wide range of other policy decisions. While organizations such as the US National Academy of Sciences had long been engaged in writing reports, few of these were influential, since they were too often formulated in a closed manner, using cherry-picked experts aiming at an unbalanced outcome. In the absence of a credible assessment process, a book by a single skillful individual might be the next best thing, if it could be widely respected for its science and also widely read. Carson was uniquely qualified both to perform a one-woman assessment as a well-informed scientist who could use her popularity and singular talent as a nature writer to ensure its wide review and exposure.

The Long Island citizen's petition was ultimately denied after a series of appeals took the case all the way to the Supreme Court by 1960. By then Carson had begun to work on the book that would become *Silent Spring*. Her mother Maria's death in December of 1958 left Carson as sole caretaker of her great-nephew, Roger. Moreover, Carson suffered from various illnesses, including a recurrence of breast cancer, requiring a radical mastectomy in early 1960. Although doctors declared the operation a success, by the end of the year Carson learned that the opposite was true, and worse yet, the cancer had metastasized. The radiation treatments that followed took a heavy toll on her energy, but she was determined to keep the news of her cancer private, lest the chemical companies use it to question her motivations. She forged ahead with the book.[28]

Finally, the bulk of the draft manuscript was ready, and it was sent to the editor-in-chief of the *New Yorker*, William Shawn. Shawn called Carson to declare the book a "brilliant achievement, . . . full of beauty and loveliness and depth of feeling." The magazine serialized several chapters, beginning in June of 1962, about three months in advance of the book's publication. The articles attracted considerable popular interest and paved the way for another best seller. They also ignited a firestorm of angry protests by the chemical companies.

On July 22, 1962, the *New York Times* reported that "the 300 million-dollar pesticides industry has been highly irritated by a quiet woman author whose previous works on science have been praised for the beauty and precision of the writing. . . . some agricultural chemical concerns have set their scientists to analyzing Miss Carson's work line by line; . . . statements are being drafted and counter-attacks plotted." But like other attempts to discredit Carson, the quotations from industry failed to identify specific scientific errors and argued rather that Carson's views were one-sided because she left out "the enormous benefits in increased food production and decreased occurrence of disease that have accrued from . . . modern pesticides."[29]

Three key national events dovetailed to set the stage for public attention just in time for the book's appearance. First, the Thanksgiving holiday of 1959 was rocked by the removal of America's beloved cranberries from grocery store shelves because a pesticide (aminotriazole) that provoked cancer in laboratory tests on rats was found in the fruit. Because this carcinogen was not just a residue on the fruit's skin but within it, the FDA used its authority to remove the crop from stores.[30] Next, the radioactive isotope strontium-90 appeared in the milk supply in late 1959, despite the phaseout of American nuclear testing in 1958, provoking widespread public concern. Carson skillfully linked the heightened public fear of nuclear fallout that had been widespread since World War II to a new and dangerous image: chemical fallout of pesticides.[31]

In September of 1960, FDA medical officer Frances Kelsey received an application for review of the new drug thalidomide, which was already being prescribed widely in Europe and else-

"I had just come to terms with fallout, and along comes Rachel Carson."

Cartoon by Herbert Goldberg, published in the *Saturday Review*, November 10, 1962.

where for the treatment of sleeplessness in pregnant women. She repeatedly stood up to the company and sent the application back, deeming the safety testing inadequate for approval. By 1961, increasing numbers of children born with flipper-like arms or legs were linked to the drug, particularly in Germany, and the horrible consequences of poor testing combined with lax regulation abroad became clear.[32] Thanks to Dr. Kelsey, the United States largely averted the tragedy that led to ten thousand malformed children in other countries.[33] The publicity surrounding each of these events successively deepened public suspicions about the rapid pace of production and distribution of new drugs and new chemicals.

Silent Spring was published by Houghton Mifflin in September of 1962. The treatise began with a vivid and allegorical chapter that explains its title by describing a fictional town stricken by a blight that killed farm animals, sickened farmers, and eliminated the wild birds, leaving the spring season devoid of its normally lovely morning birdsong. Although Carson carefully stated that "this town does not actually exist, ... (but) every one of these disasters has actually happened somewhere,"[34] the chapter would become a magnet for criticism by those eager to proclaim the work as insufficiently scientific.

Carson followed this dramatic beginning with a description of the rapidly growing use of the chemicals, intermixing facts about insecticides with her love of nature and concern about the ability of humans to destroy it. The book's literary beauty and evocative language was Carson's tool for presenting her readers with science that would engage rather than bore them. Her juxtaposition of carefully crafted and thoroughly researched scientific information and her feelings and philosophy regarding the planet, offered in her lyrical writing style, exposed targets for the critics to attack. But these same qualities quickly propelled *Silent Spring* to the spot of number one national best seller.

Carson knew that the greatest level of public concern wouldn't come from the devastating impacts on the natural world. Not everyone loved those treasures with the same intensity as she. But everyone cared deeply about impacts on that most personal of all our riches, our own health and that of human children. The book covers deaths and illnesses among "spraymen" along with the specter of unproven but possible cancers and chronic diseases. The research on human impacts was ongoing, and Carson wisely stopped short of directly overstepping what science could support. She edited and reedited the draft, carefully adding language about uncertainties.

Carson also recognized that perceptions of risk depend not only upon knowledge of facts but also on personal values. And she knew that values regarding how much to be concerned about uncertain risks would be essential factors in the reception of her book and in the future of pesticide policy. Throughout the book, Carson detailed influences on animals and left much of the key judgment on risks to human health to the reader's own values,[35] writing that "no one knows whether the same effect will be seen in human beings, yet this chemical has been sprayed from airplanes over suburban areas and farmlands." She described how DDT accumulates in fat and is passed inexorably up the food chain in larger and larger quantities, and she cited evidence that insecticide residues had been found in human milk samples as well as in the liver, kidneys, and elsewhere, remarking that even typical diets could lead to long-term poisoning. "There has been no such parallel situation in medical history. No one yet knows what the ultimate consequences may be," she said. And she summarized a series of stunning events that revealed the scope and nature of the growing damage to the environment due to use of persistent pesticides.[36]

When the spruce budworm threatened trees along New Brunswick's Miramachi River in 1953, as the insect had done about every thirty-five years, the Canadian government decided not to let nature take its course as in previous upsurges of the worm. Instead, planes rained down DDT. Within two days, dead trout and salmon lined the riverbanks. In all sprayed streams, young salmon became scarce, and fishermen bemoaned the lowest catch in many years.

Beginning in 1955, the town of Sheldon, Illinois, experienced an alarming incident aimed at the Japanese beetle, in which the USDA and the Illinois agriculture department sprayed extensively with dieldrin, another compound even more dangerous than DDT. Along with the poisoned and dying beetle grubs, dead earthworms soon surfaced, leading to massive bird losses: robins, starlings, meadowlarks, pheasants, and more. DDT also decimated ground squirrels, rabbits, and farm cats. Nor were the domestic animals spared: the farmers lamented the loss of their sheep. Carson didn't stop with the facts of the case, going on to comment that "incidents like the eastern Illinois spraying raise a question that is not only scientific but moral. The question is whether any civilization can wage relentless war on life without destroying itself, and without losing the right to be called civilized. . . . By acquiescing in an act that can cause such suffering in a living creature, who among us is not diminished as a human being?"[37]

Carson surely had many concerns that she did not present in *Silent Spring* because of a lack of sufficient scientific evidence. But she did include a speculative section on the strange drop in the population of the US national symbol, the bald eagle: "The facts suggest that something is at work in the eagle's environment which has virtually destroyed its ability to reproduce. What this may be is not yet definitely known, but there is some evidence that insecticides are responsible." Among the evidence she offered were the observations of a former banker, Charles Broley, who retired to Florida and took up the hobby of banding baby eaglets in their nests, using a rope ladder placed by a slingshot to climb up to the nests.[38] In addition to providing valuable information about migration as the banded birds were recovered later, Broley's 1944 survey documented that he had banded 128 eaglets in the stretch of Florida coast he covered. But by 1947, the production of young birds had begun dropping. Some nests went eggless and others had eggs that didn't hatch. By 1957 only 43 nests were occupied and just 8 eaglets were produced. In 1958, Broley found only 3.[39] At this rate, the national bird was headed for extinction.

Along with Broley's citizen-scientist data, Carson described the work of professionals and academics. Studies by officials at

Charles L. Broley carrying his rope ladder for climbing trees to band eagles in Tampa, Florida. 1948. The slingshot was used to position the ropes.

the Natural Resources Council of Illinois noted a drop in numbers of immature eagles, and observations at the eagle sanctuary on Mount Johnson Island in Pennsylvania revealed a cessation of egg laying and production of young.[40]

Broley was not the only citizen scientist becoming uneasy with changes he was witnessing. Noticeable falls in population of many different species alarmed birdwatchers, particularly the members of the Audubon Society around the country. Citizens in other countries noted similarly disturbing changes in wildlife.

In the spring of 1960, what Carson described as a "deluge" of reports of dead birds in Britain reached the British Trust for Ornithology, the Royal Society for the Protection of Birds, and the Game Birds Association. There were also reports of widespread deaths of British foxes. And in France and Belgium, the numbers of partridges declined.

Carson dedicated several chapters of *Silent Spring* to the available evidence regarding the risks of the new chemicals for human health, drawing on extensive interviews she conducted with medical experts. She detailed the lack of knowledge on the impacts of cumulative buildup of chemicals in the body, and described isolated incidents of disease and illness in greenhouse and farm workers, including cases of leukemia. While she was careful to avoid stating that such examples constituted proof, she opened herself up to the reasonable criticism that anecdotes don't make good science.

Along with discussion of the dangers of pesticides, Carson outlined her view of alternative methods for dealing with pests, particularly biological controls. She described the successful eradication of the screw worm in Florida, in which sterilized insects were reared in the laboratory and released, interfering with procreation and resulting in near total eradication of the pest in 1959. She also considered the use of natural enemies and of bacterial methods.[41]

Silent Spring became the October 1962 book-of-the-month club selection. Carson wrote to a friend that "this, added to other things, . . . will carry it to farms and hamlets all over the country that don't know what a bookstore looks like."[42] A whirlwind of lectures and speeches at meetings of the Audubon Society, the National Parks Association, the National Council of Women of the United States, and elsewhere followed publication of the book. The formal review in the *New York Times* for September of 1962 was written by two distinguished academics and nature writers who highlighted that *"Silent Spring* is similar in only one regard to Miss Carson's earlier books. . . . In it she deals once more, in an accurate, yet popularly written narrative, with the relation of life to environment. Her book is a cry to the reading public to help curb private and public programs which by use of poisons will end by destroying life on earth." The review goes on to say that the book is "so one-sided that it encourages argument, although little can be done to refute Miss Carson's carefully documented statements."[43]

The avalanche of accolades was matched by a torrent of attacks. Some of these derided Carson because of her career as a writer rather than as an academic scientist writing for peer-reviewed publications. Others maligned her for having the temerity to write a book about science that was full of lyrical rather than turgid prose and understandable to the public—clearly such a book could not possibly be considered authoritative and credible. Several stressed that the mix of declared love of nature and opinions about living in harmony with it presented along with scientific information automatically rendered the science questionable. This was just not the way of standard scientific books, and so could not possibly have scientific merit.

In this era long before the feminist movement, long before the "me too" revolution, the language and tone of misogyny that dominates many of the negative remarks about Carson and her work is stunning. A review of *Silent Spring* in *Chemical and Engineering News* sported the now-shocking title "Silence Miss Carson," and went on to doubt whether "many readers can bear to wade through its high-pitched sequences of anxieties."[44] *Time* magazine titled its review "Pesticides: The Price for Progress" and referred to the work as a "hysterically overemphatic ... emotional and inaccurate outburst."[45] Misogyny certainly persists in today's scientific circles and, while blatant wording still occurs, it is arguably less frequent.

While recognizing possible dangers from "misuse," many of Carson's critics were technological optimists who viewed pesticides as one of the countless, glorious scientific successes that enriched modern life. The 1962 *Time* review declared that "Chemical insecticides are now a necessary part of modern U.S. agriculture, whose near-miraculous efficiency has turned the ancient tragedy of recurrent famine into the biologically happy problem of what to do with food surpluses." The review then quotes a professional entomologist at the Illinois Agricultural Experiment Station: "If we in North America were to adopt a policy of 'Let nature take its course,' as some individuals thoughtlessly advocate, it is possible that these would-be experts would find disposing of the 200 million surplus human beings even more perplexing than the disposition of America's current corn, cotton and wheat surpluses."[46] Such criticism skillfully attacks Carson not for what she actually said, but for imagined extrapolations that were the opposite of her own careful statements, such as "It is not my contention that chemical insecticides must never be used. I do contend that we have put poisonous and biologically potent chemicals indiscriminately into the hands of persons largely or wholly ignorant of their potentials for harm."[47]

Richard Garwin, a highly respected physicist and recipient of the US National Medal of Science as well as the Presidential Medal of Freedom, sent me an email in 2019. He recalled standing at a photocopier making at least eighteen copies of Carson's *New Yorker* essay to take into an upcoming President's Science

Advisory Council meeting in 1962. The council "considered many more topics than it could add to its working agenda" according to Garwin, and nuclear and space issues dominated the group's work. Garwin's plea for the council to take on the pesticide issue would have to be weighed along with others.

The council's chair, who served as the president's science advisor and considered Garwin's *New Yorker* article photocopies, was Jerome Wiesner, an electrical engineering professor at MIT. Wiesner began to prepare for the potential storm that *Silent Spring* might unleash for President John F. Kennedy. Wiesner received a great deal of mail from members of the public expressing concerns about the use of pesticides, as well as a letter from Thomas H. Jukes, a well-known biologist and director of biochemistry at the American Cynamid chemical company, stating that "In our opinion, the main problem is Miss Carson."[48]

But Wiesner had already gotten out ahead of the critics. He decided that instead of leaving the analysis of pesticides and *Silent Spring* in the hands of the federal agencies, the council would carry out its own study, under its Life Sciences Panel. The panel set to work examining the information provided by the federal agencies, along with data from the World Health Organization, the British Government, and the American Medical Association. The Manufacturing Chemists' Association also represented the views of the pesticide industry to the panel. And the panel invited Rachel Carson to come in for an informal meeting. As Carson described it to a friend, "It was not a command performance, but just come if I'd like to." Carson spent nearly the entire day of January 26, 1963, with the panel. The council put out a draft report for comment in February 1963, which drew a torrent of industry criticism, as well as a warning from the USDA that the report could "profoundly damage US agriculture . . . and lead to a breakdown in public confidence . . . (in the) safety and whole-someness of our food supply."[49]

The council continued to gather information, including one set of new data from Great Britain that raised the issue of whether USDA should withdraw registration of dieldrin, a pesticide that accumulated in human tissues and seemed even more toxic than DDT. At a council meeting on the chemical in March of 1963, USDA and FDA officials admitted, on the basis of published studies, that dieldrin apparently could cause tumors, but they presented the remarkably tone-deaf rebuttal that "we need more data before we can decide whether dieldrin tumors are cancerous or not." Council members came away from the meeting more concerned than ever. Their report was widely viewed as an affirmation of Carson's main points and effectively quelled much of the most virulent criticism of *Silent Spring*.[50] When a group of leading scientists of such high stature have agreed with the essential arguments being put forward, it leaves little room for further scientific debate.

Even before the publication of *Silent Spring*, CBS television had contacted Carson about doing a segment for a popular news analysis show. *CBS Reports* was the early 1960s predecessor of *Sixty Minutes*, a program regularly watched by millions since its debut in 1968. Television filming and interviews frequently demand intense concentration and energy from the subject, and tire even the healthiest of scientists. By then Carson's cancer treatments were taking a heavy toll, and the producer noticed that she was far from well. Noted reporter Eric Sevareid came to interview Carson at her home in Silver Spring in December of 1962. She handled Sevareid's questions calmly and logically, and the program depicted her as an articulate spokesperson for science and nature who was bravely facing down a looming threat. Although she somehow managed to keep up a strong appearance in the film, she'd been further weakened by more radiation treatments, enough that the interview team noticed. CBS recognized the need to get the show on the air as soon as possible, because, in Sevareid's words, "you've got a dead leading lady." The program was broadcast on April 3, 1963, despite a last-minute campaign against it, apparently by the chemical industry. Three of the five largest commercial sponsors withdrew days before the broadcast, including the makers of Lysol and of Ralston Purina. The network estimated that the program was seen by roughly 10% of the adult US population.[51]

A month later on May 15, 1963, President Kennedy released the President's Science Advisory Council's report. It was widely viewed as a high-power confirmation of Carson's *Silent Spring*, mirroring the content of her book in many of its component chapters and including the clear recommendation that "Elimination of the use of persistent toxic pesticides should be the goal."[52] Sevareid commented after the council's ringing endorsement that Carson had achieved her goal to "build a fire under the government."[53]

Hearings began in the Senate the very next day, providing yet another forum for evaluation and public engagement. Carson testified on June 4, 1963. Senator Gruening of Alaska compared *Silent Spring* with *Uncle Tom's Cabin* and predicted that Carson's book would similarly change the world.[54] The hearings were widely reported in newspapers and on television's nightly news broadcasts, once again broadening Carson's reach. By now, Carson's *Silent Spring* had reached the public through every branch of media of its day: the book itself, television, magazines, and newspapers.

Nearly all of the members of Kennedy's President's Science Advisory Council were part of the generation of American scientists whose lives were deeply intertwined with World War II,

Rachel Carson testifying to the Senate on June 4, 1963. Carson is wearing a wig in this photograph due to the loss of her hair from radiation treatments for cancer.

particularly the legacy of the atomic bomb. Wiesner, Garwin, and other members had worked at Los Alamos helping to develop the terrifying weapon. More than any previous era, that period endowed science and scientists with a profound understanding that their innovations could have devastating consequences of previously unimagined scope. For nearly all of them, the experience also led to moral turmoil, as they wrestled with the anguish of how their personal actions contributed to the deaths of at least a hundred thousand innocent Japanese citizens.

By September of 1963, Carson was experiencing great difficulty walking. In April of 1964, she succumbed to the cancer that she had worked to keep secret from all but her closest friends and family until her death.[55] Despite *Silent Spring* and the eloquence of Rachel Carson, despite Garwin, Wiesner, the President's Science Advisory Council report, the *CBS Reports* broadcast, congressional hearings, and a widespread public outcry, it would take seven more years after Rachel Carson's death before DDT and other persistent pesticides would be banned in the United States. The only national law passed in the immediate aftermath of *Silent Spring* was the elimination of registration under protest, a seemingly small but nonetheless significant step.[56] US agencies were also forced to share more information with each other, and important research to resolve uncertainties regarding wildlife and human health effects greatly increased, guided by the key gaps in knowledge enumerated in the council's report.

The critics of Carson's work increasingly turned to emphasizing these uncertainties, highlighting the need for more research before, in their view, they or the government should take any action. They also trumpeted the value of pesticides for both food and human health, including control of malaria. One commentary in the widely read *Saturday Evening Post* in September of 1963 belittled Carson's mention of the death of cats in Java after anti-malarial spraying, noting the "numberless Javanese men, women, and children who had previously suffered and died of malaria" and "thanks to a woman named Rachel Carson a big fuss has been stirred up to scare the American public out of its wits."[57] Robert H. White-Stevens, a frequent Carson critic and scientist with the American Cynamid Corporation, exhorted industry colleagues and farmers to "tell the urban peoples in a thousand places and a thousand ways what scientific agriculture . . . has meant to their health, their welfare, and their standard of living, . . . that DDT has saved as many lives over the past fifteen years as all the wonder drugs combined, . . . [and] that agricultural science . . . has for the first time in man's long struggle against want procured the means to banish hunger from the earth in our time."[58] The critics attacked the idea that people should feel personally threatened, and questioned whether other approaches to ensuring adequate food for humanity were practical.

Even a number of the most distinguished wildlife scientists in the United States, including for example ornithologist Joseph Hickey of the University of Wisconsin, were initially skeptical of Carson's work on the scope of impacts on wildlife.[59] In short, in the early 1960s, the public and policymakers were concerned yet hesitant, because the pesticide issue was not obviously personal to enough people, because broad effects on wildlife were not yet fully perceptible to enough of them, nor to a strong enough majority of experts, and because the industry vigorously promoted the fear that alternatives simply could not be practical.

Nevertheless, the soft power of Rachel Carson's book and its aftermath was profound. Widespread public concern following *Silent Spring* motivated some policymakers to use their authority to the extent allowed by law. Secretary of the Interior Stewart Udall directed his department to avoid using DDT, chlordane, dieldrin, and endrin on the 550 million acres of public lands they administered, unless no substitutes were possible.[60] While it is difficult to be certain how many individual farmers or communities chose to stop using the pesticides of their own accord without legal restrictions, records of US national sales of DDT suggest the power of Carson's persuasion: about thirty thousand metric tons of DDT were sold yearly from 1956 through 1963, but only about twenty-two thousand from 1963 to 1967, the first five years following publication of *Silent Spring*. By 1970, no national bans had yet been passed, but US usage had dropped further, to about ten thousand metric tons.[61] Carson surely started this precipitous downward trend in sales, but momentum built after her death because of other scientists, citizens, and the media.

Scientists continued to do what they do best: develop the needed information to close the key research gaps identified by the President's Science and Advisory Council. They also shared their findings at conferences to pool knowledge, not only across different research groups but even across continents. And of all that research and data-gathering, none had more impact on public understanding and policy action than the rapid collection of convincing evidence that continuing the use of pesticides could indeed lead to the extinction of falcons, bald eagles, and other raptors.

The peregrine falcon is a small but spectacular member of the raptor family, the world's fastest creature that can fly at speeds exceeding two hundred miles per hour when descending on its prey. During the Twelfth International Ornithological Congress in 1962, the assertion circulated that no peregrine in the US Northeast had raised young in the past year. Joseph Hickey initially dismissed it as a rumor. As a peregrine expert, he knew that the birds return year after year to the same nest, that the best and most remote nests pass between generations, and that peregrine populations had a history of stable numbers dating back more than a century in the United States.[62] But like all good scientists, he would shortly find his views changing as the evidence mounted.

Across the ocean in Britain, something strange was indeed happening with the peregrines, and the evidence was not just anecdotal. Derek Ratcliffe was a scientist for the Nature Conservancy Council in the United Kingdom. Over 170 volunteers helped him collect data on peregrines by climbing up to their nests. In a paper published a year after Carson's book in 1963, Ratcliffe documented a remarkable falloff of successful peregrine breeding even in the single year between 1961 and 1962. For example, in fairly remote parts of Scotland, only half the nests were even occupied in these years. And only 15% of the birds successfully raised broods the first year, dropping to only 4% in the second.[63] Similar reports came in from France, Finland, Sweden, and Germany. Ratcliffe considered possible factors that might account for this disturbing finding, such as changes in food, climatic shifts, or disease, but found it difficult to ascribe the declines over such widespread areas to those effects. He also noted that Carson's evidence for pesticide residue buildup in food chains in North America corresponded with the peregrine's difficulty to reproduce: "The wave-like spread northwards of decline [in Britain] coincides closely with the pattern of use of organic pesticides, both geographically and in time."[64]

Hickey knew what he had to do. He set out to test Ratcliffe's hypothesis by evaluating whether the same things were happening in North America. He sent two students on a trip of fourteen thousand miles to check peregrine nests from the state of Georgia to Nova Scotia, following a route with 133 known nests that he had surveyed himself in 1939–1940. The chilling results came when the students reported that not a single peregrine fledgling was found anywhere in 1964.[65]

A startled Hickey decided to convene an international conference himself in Madison, Wisconsin, in 1965, bringing in his colleagues from all over the world to compare their findings. A march toward peregrine extinction was underway nearly everywhere, with reductions in young of 80 to 90%. The group of scientists identified a "common pattern" where the falcons first laid infertile eggs, then thin-shelled ones that generally broke, and then no eggs at all. The species was headed for oblivion unless something changed.[66]

In the United Kingdom, Ratcliffe came up with a novel idea for further analyzing the problem. He collected eggshells from museums and private collections dating back to 1900 from across Britain. These revealed a marked drop in eggshell thickness (as measured by relative weight after accounting for size) after about 1945, not only for the peregrine falcon, but also for another species, the sparrowhawk. The same behavior was found in US ospreys and peregrines on the East Coast. And in Florida, the state where Charles Broley found the bald eagle population plunging, the thinning of eagle shells was also documented. Moreover, chemical analysis at many of these sites revealed that the thinner the eggshell, the higher the amounts of the products produced within the body as it metabolized DDT, namely DDE and DDD.[67]

These data, while concerning, did not represent unassailable proof that the cause of the extraordinary changes was insecticides. Extraordinary scientific results require extraordinary evidence, including not just observations and correlations but controlled lab studies and clear chemical or physical mechanisms. So another group of wildlife biologists came up with a plan to study the effect of insecticides on sparrowhawks, a species that could be raised in captivity. This team included Lucille Stickel, whose very existence as a female scientist the *New York Times* had reported as a charming oddity about two decades earlier. Compared to a control group fed insecticide-free food, Stickel and her colleagues demonstrated thinning of the shells for birds fed DDT and dieldrin.[68] The biochemical mechanism of the effect was identified in 1967 by another researcher, who connected DDT and similar compounds to reduced production of steroids, which control the mobilization of calcium that female birds need to make an eggshell.[69] The effect on raptors was so stark because they were at the very top of their contaminated food chains, eating prey that had in turn eaten smaller animals that had eaten even smaller animals or plants, with each step concentrating the pesticide to higher and higher concentrations. By the end of the 1960s, the scientific evidence that DDT as well as related pesticides were threatening the existence of eagles, falcons, and many other birds of prey overwhelmed any doubt that Carson's speculative inclusion of the issue in *Silent Spring* was the right call.

Ornithologists were not the only scientists intent on the trail of DDT. Aquatic biologists also rose to the challenge. Research concluded that persistent pesticides at very low concentrations decreased the survival rates of young fish.[70] Science erased any lingering arguments that pesticides were merely a local problem, and they had shown that it pervaded national waters.

Public concern continued to build as reports of the science came out in the media, but industry and the USDA steadfastly pushed back on US national policy change, mainly by pounding the drum of how essential these compounds were. Like most such assertions about environmental concerns, the proponents took extreme stances rather than debating the real issue of whether a more moderate approach could be practical, not an immediate all-out ban. The Secretary of Agriculture, Earl Butz, flatly declared in 1972 that without pesticides farmers could not produce enough food for 206 million Americans. But in fact, by 1973 about half the remaining use of insecticide in America was devoted to growing cotton rather than food.[71] Growing cotton certainly had economic benefits, but it was not relevant to food production. How effective were pesticides at increasing food production compared to not using them? Research on the effectiveness of pesticides, fungicides, and herbicides can be carried out using, for example, two test plots, only one of which is treated with pesticides. Studies in 1974–1978 estimated that the use of the chemicals saved about 10–15% of national food crops, again implying significant economic benefits (about $8.7 billion in 1978) but not supporting claims of a food supply catastrophe without them.[72] Further, such estimates of economic benefits of potential pollutants then (and now) disregarded "external costs" that act to reduce the real gains involved. They ignore, for example, the cost of treatment for human pesticide poisonings and illness, and livestock losses. They also ignore the incalculable: the loss of human life, the wholesale reduction of wildlife, and the value attached to nature and a healthy human environment. The tide began to turn when people became concerned enough to organize and learned to make use of a potent weapon to enact change: the law.

A group of private citizens, mostly from Long Island, New York, came together in the mid-1960s to begin holding monthly meetings in each other's homes to informally discuss a range of environmental concerns. The group included scientists from local institutions, including Brookhaven National Laboratory and the State University of New York at Stony Brook, among them biologist Charles Wurster, who had already published scholarly work on the influence of DDT on birds. It also included a few high school students and a high school biology teacher, and would ultimately include Victor Yannacone Jr., an attorney. They initially dubbed themselves the Brookhaven Town Natural Resources Committee, but the name and the organization would soon evolve into one of the most powerful and influential environmental groups in the world, the Environmental Defense Fund (EDF), which drew its name from the Legal Defense Fund of the National Association for the Advancement of Colored People.[73] That evolution centered on a remarkably clever grassroots campaign against the misuse of DDT that soon became the group's ardent focus.

Since 1947, the local county's mosquito-control commission had been generously spraying DDT in local marshes. The

Brookhaven citizens had been trying to stop this practice for years without success, so in 1966 they delegated Wurster to write a letter to the editor of the local newspaper about the risks associated with the practice. They had written letter after letter before, to no avail, but this one was special. Its appearance prompted a phone call from the lawyer Yannacone, who together with his wife Carol was working on a lawsuit against the county for allegedly flushing the tank of their DDT spray truck into a lake located about five miles away from the town of Brookhaven. Carol had grown up in the area and was concerned about recent fish kills—the recurring sight of hundreds of dead and dying fish that she had never witnessed as a child.[74]

Yannacone joined the Brookhaven committee at its next meeting and introduced them to his watchwords: "Sue the bastards." The excited group gathered scientific reprints and wrote affidavits on the destruction of wildlife that they had witnessed, and Yannacone filed the papers at the State Supreme Court to request an injunction against the commission. Newspapers reported on the lawsuit, spreading the word to other members of the public. Just three months after Wurster's letter brought Yannacone into the fold, a temporary injunction was issued, enjoining the commission from using DDT on the marshes. The group was stunned. After nearly two decades of dedicated struggle with other approaches, the lawsuit had stopped the spraying in weeks. Wurster wrote that "letters to politicians received form letters in reply, . . . (but with the lawsuit) we had grabbed the mosquito commission by the tail."[75]

The case went to court in November of 1966, and the judge did not yet rule but kept the injunction in place through the winter and into 1967. With no option to use DDT, the mosquito commission used a different insecticide, Sevin. With an environmental lifetime of only a few weeks, this pesticide's behavior would be starkly different from DDT, with little accumulation in the environment or spreading beyond its point of use. After this gain of experience with alternatives, DDT was voluntarily prohibited by the county in 1967 and would never be used there again. All this occurred before the judge rendered a decision throwing Yannacone's lawsuit out of court in December 1967, on the grounds that it was not up to the court to set guidelines for the county's insect control; rather, any ban must occur via the legislature. But by this time the final ruling was moot, and the group had learned an incredible lesson: they had "won while losing."[76] The process of the lawsuit had forced the county to give up DDT—despite the fact that the suit was eventually thrown out!

The group next entrained other members by speeches at the national Audubon Society convention, and ultimately ten of them signed a certificate of incorporation of the EDF, with an explicit goal to "marry science and the law to defend the environment in the courts," according to Wurster. Wurster would become the EDF's first chief scientist.[77] By the 2020s, the EDF would grow to include over three million members.[78]

The EDF proceeded to introduce lawsuits in Michigan, including one to stop nine towns from using DDT to control Dutch elm disease. None of the towns contested the lawsuit. They all stipulated that they would no longer use DDT rather than deal with the expense, negative publicity, and time of a trial, and the suits were therefore dismissed, never making it to court. EDF again triumphed without winning, and neither side had a chance to figure out if they would have lost. The EDF immediately named fifty-six more towns that were still using DDT for various control strategies, and they all stopped using it too.[79] But while they had succeeded in stopping local DDT use in dozens of towns, they had not had the chance to fully present their evidence in court, and hence in the more important and much broader court of public information and opinion.

The members of the EDF set their sights higher. Rather than going case by case, they wanted to expose the nationwide risks associated with pesticides. They went after dieldrin, a similar chemical already shown to be even more dangerous than DDT. But when they tried to introduce a suit against nurserymen using dieldrin against the Japanese beetle in Michigan's forestry country, they were thrown out of court as having no standing in that state. They decided to try again in Wisconsin, on the other side of Lake Michigan, on the grounds that the spraying would pollute the lake and hence both states. They were thrown out again as having no standing and without impacting the dieldrin use. After all, they were not even residents of Wisconsin. It was clear how to solve that problem. They allied with a group of Wisconsin conservationists who were trying to halt DDT spraying for Dutch elm disease in Milwaukee. The group filed an initial complaint with the Wisconsin Department of Natural Resources. But as in New York, the county agreed not to use DDT in a Department of Natural Resources hearing before the case went into formal court proceedings, rendering it moot.

The chief hearing examiner was perplexed when the EDF's members bemoaned the outcome. After all, it seemed that they had achieved their objective, so why weren't they happy? He then learned that their goal was far loftier: Yannacone explained that EDF was setting its sights on finding a high-profile public forum to fully present the scientific evidence and get a judgment that would influence DDT use not just town by town, but at a much broader level.

The hearing examiner pointed out that they had picked the wrong legal route. He revealed that Wisconsin had just the legal mechanism they really needed: any Wisconsin citizen could ask any state government department for a ruling on the applicability of a particular set of facts to any rule enforced by the department. The agency would have to evaluate the request and decide whether to hold hearings. Then each side would present their case, and a declaratory ruling would ultimately result. Unlike the courts' procedure, the process had to go to completion once it had begun. The EDF could, for example, request a hearing to determine whether DDT was a water pollutant, citing Wisconsin's water-quality standards. There they could present as much evidence as they liked about the impacts on wildlife and potential risks to humans. The industry would have a chance to rebut, and the hearing would have to go through to a ruling. This obscure law could be the answer to EDF's dream. Just ten days later, the citizens of the Wisconsin conservation group, backed by the EDF, asked the state department of natural resources for a declaratory ruling on whether DDT was a water pollutant—after all, it was killing not only birds but also fish. It also might be a threat to humans, and alternatives were available. Under Wisconsin law, the agency need not grant every requestor a hearing, but public concern over DDT had reached sufficient intensity that they decided it was appropriate to do so. DDT would be on trial in front of the entire nation, beginning in December of 1968 in the city of Madison. As they presented their evidence, they would have a chance to argue that the DDT released in Wisconsin could have impacts in other states (just like those remote Scottish peregrine falcons, whose poisoning was not local), making the case one of national interest.[80]

The *New York Times* called the hearing "the first forum in which the nation's scientific community has been able to meet the [chemical] manufacturers in a face to face confrontation that can be carried to a decision," and the hearing could "affect the use of pesticides in every state." Madison's *Capital Times* called the hearing a showdown between David and Goliath: Goliath big, big-moneyed and silk-suited; David "passionate but poorly funded." Cameras from the news networks provided regular national coverage throughout the proceedings. The passion of the private citizens who described their own observations to reporters outside the hearing room, together with the testimony of cautious scientists made for good news stories. And the appearance of one of Wisconsin's own senators, Gaylord Nelson, attracted a level of media coverage that he had been utterly unable to muster by his formal speeches on the senate floor about DDT. Nelson's testimony led to banner headlines across the nation.[81]

The petitioners brought in a series of articulate researchers who painstakingly presented the scientific case against DDT, including Stickel, who had demonstrated thinning of the shells for birds fed DDT; ornithologist Hickey; Wallace, who had sounded the alarm on the robin deaths; as well as Wurster.[82] Wurster was characterized as particularly impressive during every cross-examination by the industry lawyers because of his nearly photographic memory. It was a matter of great surprise to Wurster that a glance at a television set during dinner one evening confronted him with his own image, delivering the day's testimony.[83] And this was not just any TV news show—it was the *CBS Evening News* with Walter Cronkite, the leading national program watched at dinner tables all across 1960s America. Evidence focused on the long lifetime of DDT, its dispersion all over the world, and its impacts on birds and fish. The hearing also included a point that could not fail to attract public concern: that ongoing residues of DDT were present not only in human tissues but also in mother's milk. What could possibly be more personal?

Industry bungled the choice of its primary lawyer, picking a skillful lobbyist with no experience in the challenges of cross-examination. He sought no scientific advisors and was replaced halfway through their part of the hearing. The proceedings continued into 1969. Shortly before the end of the industry's attempt at rebuttal, an unexpected event propelled the hearing to a new level of importance and prominence: the FDA banned sales of over thirty-thousand pounds of Coho salmon harvested from Lake Michigan because the fish had been found to contain more DDT than the legal limit. In addition to being a local and readily perceptible story, the crescendo of exposure propelled change. Even before the hearings ended in May of 1969, the Wisconsin state legislature began action to ban the use of DDT throughout the state.[84]

By 1970, the environmental movement was in full swing with its new tool, the courts. EDF brought a lawsuit against the USDA to ban DDT entirely nationwide. There would, of course be suits and countersuits, appeals and more appeals, not only on DDT but also on many other environmental concerns in coming years. But winning (even while sometimes losing) boosted the environmental consciousness of a nation already influenced by Rachel Carson. Environmental organizing moved from the living room to the courtroom. The EDF and other organizations gained support, took on full-time employees instead of relying only on volunteers, and grew in scope and complexity.

While the use of the legal system has become a larger and larger factor in determining how environmental issues evolve, it has also provided a convenient target for those who wish to dismantle environmental protections as too bureaucratic, expensive, or burdensome. And the link first forged by control of persistent pesticides between science, the courts, and regulation has been repeated with many other environmental issues. The power of fact in a court case has also made science and scientists even more of a prime target for industry attacks.

Concerns about pesticides mounted following *Silent Spring*, not only in the United States but also in other countries. Environmental history shows that fractured authorities dealing separately with air, water, and wildlife are very often a formula for bureaucratic sidestepping and delay, while actions occur more effectively in countries with a single national entity whose mission is the big picture of environmental preservation.

Institutions of government may seem to be an utterly boring component of environmental policy, but they are its bedrock. The Swedish public became increasingly concerned about the environment in the 1960s, prompted by *Silent Spring* as well as reports about acid rain's increasing damage to their lakes and forests. In response, the Swedish government reorganized in 1967, consolidating nature conservancy, water and air protection, outdoor recreation, and the protection of wildlife in one governing structure with integrative authorities, the Swedish Environmental Protection Agency. A two-year temporary national ban on the use of DDT in Swedish agriculture and gardening followed promptly in 1968,[85] and was shortly replaced by a permanent ban. Along with the new agency, a national tradition of environmental protection and the public- rather than private-oriented culture of a nation governed as a social democracy are among the factors that led to Sweden's swift actions.[86] And Sweden has long been largely an importer of specialty chemicals such as pesticides, with a limited domestic chemical industry to protect, unlike other nations such as the United States, the United Kingdom, and Germany. Australia was another early actor: limits and bans were enacted in several Australian states by 1968. Australia, like Arizona and Wisconsin, was heavily influenced by its dairy industry.[87] Meat, butter, and milk were Australia's major exports in the 1960s.

But within the United States, the diversity of federal agencies that were involved in regulation of pesticides slowed efforts to take any national action: with the USDA, the FDA, and the Departments of Interior and of Health, Education, and Welfare all in the mix, policy change simply could not be fast. And officials at the USDA remained closely aligned with what they perceived to be farming interests, even supporting the chemical industry side in one of the lawsuits brought by the EDF. The turnabout came in December of 1970, when the United States government underwent a major overhaul under President Richard Nixon. While focused in principle on issues of efficiency, among the actions taken was the creation of the US Environmental Protection Agency. For the first time, the United States had a single agency whose mission was to protect the environment. The regulation of DDT and other dangerous pesticides was transferred away from the USDA to EPA's large initial to-do list, along with many other issues.

William Ruckelshaus was the first director of the new agency. Ruckelshaus later reflected that "one of the factors leading to the creation of E.P.A. was the recognition that without a set of federal standards to protect public health from environmental pollution, states would continue to compete for industrial development by taking short cuts on environmental protection."[88] Ruckelshaus wisely positioned the EPA's charge to align with the personal health of the public, as a means to garner popular support for the political fights that he knew would begin as soon as his young agency started challenging polluters. He set to work on the pressing environmental health issues of the day, including air pollution and pesticides.

The EDF promptly redirected its suit to ban DDT from the USDA to the EPA. This time the lawsuit was probably warmly welcomed; indeed, it was probably a "friendly" case, forcing an action that Ruckelshaus and his agency were eager to take. In January 1971, the EDF and other environmental groups that had joined with them won a federal appeals court order seeking that the EPA cancel approval of DDT, which it immediately did. Industry responded by demanding review—namely, another round of public hearings and a scientific report by the US National Academy of Sciences. The findings of this federal examiner differed from those of the examiner in Wisconsin, with a view that DDT's benefits tended to outweigh its risks and recommending that "essential" uses be exempted from a ban, including its use on the large American cotton crop. But under the law, these were only non-binding recommendations. Ruckelshaus ignored it and banned DDT in nearly all applications, apart from medical emergencies and a few specialty crops.[89] Years of court appeals followed, and the US ban survived every challenge.

But the DDT phaseout was directed only at US domestic use and did not affect the annual exportation of more than ten million metric tons to other countries at the time. Such exportation, or "dumping," allows a domestic chemical company to continue to profit from sale of materials that would be illegal in their own country, often by selling to low-income countries. And indeed, use of DDT and its sister poisons continued for many years in farming over much of the developing world, where the imperative to improve the desperate lives of the population by producing enough food and income often eclipsed any concept of protection of the environment or human health. One of the worst of the chemicals Carson described, aldrin, continued to be used for example in South America, where it was believed to be responsible for thirteen human deaths in Brazil in 1985, after it had been banned in the United States for more than a decade.[90] In Zimbabwe, one study examined the ongoing lake pollution due to the continued use of DDT for growing cotton and corn through 1990.[91] Ironically, some of that cotton likely found its way into fashionable clothing that flowed in its turn back to the very same developed countries that had long banned the use of DDT on their own soil, but readily sold Zimbabwe the chemical.

Many lower-income countries are too poor to support the institutions and government structures that were so important in controlling the use of DDT in Sweden, the United States, and other higher-income countries. As late as 2011, for example, over a quarter of the African and Southeast Asian countries responded to a WHO questionnaire by indicating that they had no legislation at all for national or regional registration and control of pesticides.[92]

But as public understanding and scientific evidence grew, and as more and cheaper substitutes for DDT and related chemicals were developed, the United Nations took aim at the need to improve the lot of countries worldwide. In May of 2001, following decades of patient and frequent discussions among diplomats, an international United Nations agreement was adopted on Persistent Organic Pollutants (including DDT, aldrin, dieldrin, endrin, chlordane, heptachlor, and six more "dirty dozen" chemicals, most of which had been the subjects of *Silent Spring*). The convention aimed to phase out global production and use of the dirty dozen, except in limited exempted uses. Wealthier countries pledged to assist lower-income countries in transitioning to non-listed substances.[93] As of 2022, 186 nations were parties to the convention.[94]

Modern Carson critics often attack her for the decreased use of DDT to fight malaria, with some even laying the blame for the deaths of millions of children in the developing world at her feet. But to do so is to ignore what Carson herself declared: "I do not say that these insecticides should never be used." While she can be criticized for not discussing potential exemptions, neither did she argue for a complete ban.

A key exemption provided under the Stockholm Convention is the ongoing production of small amounts of DDT to combat malaria in developing nations by spraying it carefully and sparingly within homes:[95] a markedly different approach than the massive doses applied over extensive areas in the heyday of DDT's reign. There is some evidence that the mosquitoes that spread malaria have evolved resistance to DDT, and may have done so even before environmental concerns mounted.[96] But the use of DDT against malaria is still managed annually by the WHO under the Stockholm Convention, at the request of each country that may choose to do so. And while the malaria exemption remains in place for DDT, all parties to the Stockholm Convention have agreed to stop producing aldrin and dieldrin, the worst of the persistent pesticides highlighted in *Silent Spring*.[97]

China was among the last countries to cease production of DDT in 2007. The only remaining producer by the 2020s was India, which continued to sell it to other countries and to use it themselves, for control of malaria and another dreaded tropical disease, leishmaniasis. Spread mainly by sand flies, leishmaniasis produces large open skin sores that resemble leprosy. While it can be treated, there is currently no proven vaccine.[98]

Over much of sub-Saharan Africa, the use of DDT on the walls and windows of homes to combat malaria and other diseases has been phased out by governments since about 2009, in favor of specially treated individual bed nets. The nets are treated with a long-lasting insecticide (not DDT), and can protect a person for years at a cost of about only about $2. Free nets are widely distributed across the globe by governments and charitable organizations, including major contributions from the United States, Britain, the Red Cross and Red Crescent, and the Bill and

Melinda Gates Foundation, which has made ridding the world of malaria one of its targets. Worldwide DDT production decreased from over eighty thousand metric tons in the United States alone in 1963 to less than three thousand worldwide by 2014.[99] The phaseout has led to a measurable drop in the DDT levels in human tissues worldwide, including in people in the developing world.[100] And the populations of bald eagles, peregrine falcons, and other raptors have displayed widespread rebounds.

The long persistence of DDT and its sister compounds means that there are still some places that remain contaminated and will continue to be for some time to come, particularly the sites of former chemical companies. DDT continues to poison robins and other birds more than forty years after its ban—if they happen to feed at the former location of Velsicol chemical in Michigan, which is now a Superfund site.[101] There is also evidence that some DDT is still slowly released today into remote places, from reservoirs of glacial ice. Even in otherwise-pristine Antarctica, researchers have detected legacy DDT that was produced many decades ago in the bodies of Adelie penguins (albeit at levels too low to cause measurable harm to the birds).[102]

We continue to use pesticides, but they are a very different breed of compounds, and might be considered very distant cousins of the original sisterhood of persistent pesticides, DDT, aldrin, dieldrin, and others. While the new compounds are generally lethal to their target bugs, fungi, or unwanted vegetation, they persist only a short time in the environment, lasting long enough to carry out their deadly tasks but not enough to travel very far or build up from year to year. This does not mean that these new cousin chemicals are totally safe. Some citizens and environmental groups argue that a number of them pose dangers to wildlife and people in their own right. But they will not be found in the eggs of birds in remote parts of Europe or North

America, much less in the tissues of penguins of Antarctica. They represent local and temporary hazards, rather than the broadside, slowly growing assault across so much of nature that the persistent pesticides represented.

The family of short-lived pesticides, herbicides, and fungicides in use today receives ongoing scrutiny. The neonicitinoid chemicals used in agriculture have been implicated in the surprising decline of beneficial bees in much of the world and have been banned in several countries,[103] including those in the European Union. Because the European Union has a large population and represents a substantial part of the world market, this policy action makes the chemicals less profitable for its now-global makers and puts strong pressure on the industry to seek alternatives. The United States has also banned some neonicitinoids, with more under review.[104] Researchers have also found a mysterious decline across the spectrum of land-based insects at more than a thousand worldwide sites, including not only agricultural areas but also protected sites. But while studies show a decline in terrestrial insect mass of about 30% over recent decades, freshwater insects are actually increasing,[105] suggesting that multiple influences are acting upon these small but essential creatures. Insects represent a critical bedrock of nearly every food web, providing nourishment for small birds, reptiles, mammals, and fish, many of which in turn feed larger organisms. While habitat degradation has been implicated in insect declines, modern pesticides are also suspected contributors.

Glyphosphate, long sold by Monsanto under the trade name Roundup, is a widely used herbicide that is under attack as a potential cause of certain human cancers, particularly non-Hodgkin's lymphoma, and continues to be debated in court.[106] The Bayer corporation acquired Monsanto shortly before these court cases developed, and their stock has plummeted. Chlorpyrifos is an organophosphate neurotoxin pesticide. Studies suggest that it poses risks to children's health, including increasing asthma and developmental delays. Its recent history reveals the back-and-forth of US environmental decision-making. It was banned from use on US crops by the EPA under the Obama administration, but the Trump administration reversed the national ban. But when the federal government fails to lead, states can sometimes step in. The states of Hawaii, California, and New York all banned chlorpyrifos. Since so many of America's fruits, vegetables, and nuts flow from California to American tables, this single super-regulator state has an outsized impact on the market and hence the viability of chlorpyrifos. Together with the ban instituted by the European Union, the loss of lucrative and large markets put strong technology-steering pressure on the manufacturers to find substitutes. And in 2022, the Biden administration banned chlorpyrifos on food crops.[107]

In many rich countries, wealthier consumers advocate organic products and the avoidance of chemicals in agriculture altogether, even if it costs more (and sometimes it doesn't). Recent decades have seen a sharp rise in the use of non-chemical methods that provide competition to the entire pesticide industry: for example, growth of integrated pest management, employing such tactics as the use of natural enemies of the undesired pests; changes in farming practices, such as crop rotation; and changes in planting and harvesting dates that reduce the need to combat pests with chemicals.

The phaseout of DDT and other persistent pesticides represents the culmination of the actions that first ignited the global environmental movement. That DDT was found in human mother's milk and in the fatty tissues of humans made the issue deeply personal, despite continuing uncertainties to this day regarding potential human health impacts. Whether DDT can cause cancer remains controversial, with some evidence for impacts decades after exposure.[108] But there is clearer evidence that aldrin and dieldrin do result in human health impacts, including cancers,[109] as Carson feared. And people just don't like the idea of man-made chemicals slowly accumulating in their own bodies, or those of their own children. The deaths of birds and fish made the issue perceptible, even to those uninterested in the natural world.

In relatively short order, a variety of practical solutions were found. Societies worldwide continue to struggle with balancing the need to feed a growing population, in the developing world in particular, with the need to avoid excessive costs and labor as well as undue chemical harm. Nations worldwide also struggle with evaluating the safety of dozens of new, potentially useful chemical agents within efficient regulatory structures. And in most cases, the corporations in the chemical industry have recognized through painful experience that continuing to make products that are shown to be harmful usually becomes a losing proposition that a smart company would avoid. They have much more to gain from a transition to making new and safer substitutes, since it is the very same companies who have the capabilities needed for manufacturing any type of chemical. They stand to make a tidy profit if they can be first and best with a safe and effective product that is cheap enough to attract a large market.

The story of the phaseout of persistent pesticides illustrates the confluence of a series of factors: first and foremost, the power of the public, as well as that of a uniquely gifted writer, Rachel Carson, who inspired them. Next came other scientists, who confirmed her fundamental points and greatly expanded the level of evidence regarding impacts on birds and fish into unassailable proof, which fueled the environmental groups who organized citizens and learned to use the power of law. The development of institutional structures enabled governments to take action individually and collectively, and skillful worldwide policymakers ultimately acted decisively in both the developed and developing worlds to safeguard both nature and ourselves.

Environmental change can seem impossible when big industries are involved whose reach permeates human life. But the fact that food and farming, a mammoth industry that could not be more fundamental to society, can and did change offers a bright beacon of hope for today's major issues.

6

Climate Change

TIME TO SEIZE THE DAY

I noticed that it was exceptionally hot for July. The relentless hot Sun beat down nearly every day, browning and shriveling the landscape all around the house that my husband and I had built in the foothills of the front range of the Colorado Rockies. Occasionally, feeble afternoon clouds struggled to form the characteristic thunderstorms that typically quench western North America's summer thirst, only to dissipate and vaporize.

And then the fires started. I watched the slurry bombers pass over—large, slow-moving planes on their way to drop chemical retardants onto a growing fire south of us. The powerful beat-beat-beat drone of heavy-lift helicopters accompanying them to drop huge buckets of water boomed across the canyons. Ash began to fall on our deck, filtering down silently like snow. As days passed, the smell of smoke became overpowering. Despite the dogged efforts of the dedicated firefighters, that fire grew until mercifully hesitating its advance just three miles south of our doorstep. It was 95°F, and some of that oppressive heat was human-caused, mainly the result of the burning of fossil fuels, particularly coal, gas, and oil. While forest management may have increased fire in some places, there's little doubt of climate's impact. Even in remote places like western Canada, where there is little to no forest management, the acres burned annually are increasing systematically as the planet warms. That's because climate change not only heats up the air but also sucks the moisture out of the vegetation, causing regional droughts and leaving the landscape primed to burst into furious flame at the slightest provocation, whether that's a lightning strike or an errant cigarette butt.

My studies of climate change had certainly convinced me of its acute dangers years ago. But, like most people in rich countries, in my heart I thought that for decades to come it would be others who would bear the suffering—the poor and the vulnerable. But in 2020, climate change suddenly became perceptible and gut-level terrifying to me personally, as western North America seemed engulfed in a plague of fire. Like the successive plagues of the Bible, the climate change plagues are varied. In the United Kingdom, the climate plague is increasing heavy rain and flooding. The plague manifests in Fiji as more powerful typhoons. In East Africa, there is even evidence that climate change is causing unusually severe infestations of locusts.

This is all frightening. But, as you might suspect at this stage in this book, I'm convinced that there is a brighter side. Now is a golden moment. If we seize the day within this decade, we can craft a better future for life on Earth. Understanding the basic science, the global politics, key economic factors, and the essential roles of the public and of technology-steering shows that the world is on the cusp of a brighter future. Just as it was for other environmental issues, success is within our grasp.

In June of 1991, Mount Pinatubo, a volcano in the Philippines, erupted in a fiery explosion of molten lava, steam, and blazing embers. Rock debris and ash rained down throughout the island nation, killing hundreds of people and rendering about a quarter of a million homeless as apartment buildings and houses collapsed under the weight of the debris. The explosive volcanic plume rose to a towering twenty-five miles above the Earth, depositing a massive load of sulfur dioxide gas directly into the stratosphere. That gas was soon oxidized and absorbed by stratospheric particles, creating a massive haze layer that spread over the entire globe within a few months. Spectacular sunsets followed for nearly a year, making the skies around the world glow red and purple even on a cloudless day.[1]

James Hansen of NASA's Goddard Institute for Space Studies has been one of the leaders in climate science for decades, and he wisely predicted that a detectable global cooling of the planet of about a half degree Celsius would result from this eruption by 1992, and would last for several years.[2] It didn't take long for global average temperatures to start falling as expected, cooling by about 0.6 degrees Celsius by the summer of the next year. Pinatubo wasn't the first big eruption to do this in human memory—among many examples, the famous "year without a summer" in 1816 followed an even bigger eruption, the Tambora volcano. Many perceived changes in climate during the year that followed. Among them was amateur meteorologist and former president Thomas Jefferson, who remarked on "the most extraordinary year of drought & cold ever known in the history of America," although he did not connect it to Tambora.[3]

The Pinatubo eruption, like other major volcanic events before it, is a tremendously instructive natural event that demonstrably affected global temperatures. Humans don't cause volcanic eruptions, of course, but we can study them to gain insight into other climate changes that we do cause. These eruptions show that Earth's global climate changes when its flows of energy are pushed into an altered state.

So what are these energy flows, and how do they work? Earth's temperature is linked to flows of both infrared and visible light. If you look at a person, or a cat, or any other warm-blooded creature using night-vision goggles in pitch darkness, you'll see them glimmering and glowing. That's because they are radiating a type of light that is characteristic of their body temperature. It's infrared light that your eyes can't see on their own, but the goggles change the infrared into light that you can see—with them on, you can even tell that a cat's ears are cooler than its face. The Sun has a temperature of thousands of degrees, so in addition to emitting in the infrared, it radiates a far more energetic kind of light, mainly visible light, across our solar system and beyond. Planet Earth has a temperature too. Our planet's temperature is largely maintained by a balance between two things: warming by the glorious visible light from our powerful Sun falling on land, soil, and oceans, versus cooling by the radiation of Earth's own infrared light back to space at its characteristic temperature. Some of that infrared light can be absorbed by clouds or greenhouse gases in the atmosphere, including carbon dioxide, and that absorption turns into heat kept on the planet instead of being lost to space.

When Pinatubo covered Earth with haze, some of the Sun's energy was reflected back out to space before it could even reach the ground, and that's what caused the volcanic global cooling. The eruption therefore provided a vivid demonstration of how changes in radiative energy flows force planetary climates to change ("radiative forcing" in climate change lingo). Pinatubo's haze layer was flushed out of the stratosphere by the atmospheric winds and rained or snowed out within a few years, and the energy balance essentially returned to its pre-Pinatubo state. In contrast, carbon dioxide has increased gradually but inexorably since the beginning of the industrial era, and the man-made increases in this powerful greenhouse gas are slowly but surely pushing our climate into a much more long-lasting altered state. Although it seems slow to a person, it's dangerously fast on the planetary time scale, and the speed of the change is a threat to the capacity of the planet and its life to adapt. To name just a few examples, sea ice losses are already causing destruction of some Alaskan native villages; sea level rise is already inundating several island nations and threatening their very existence; the death tolls of more frequent and severe heat waves are rising; droughts are increasingly sapping the ability of humans to raise crops in parts of the world and causing habitat loss. Scientists have known and predicted these and other risks for decades, and the Pinatubo eruption's evidence of the dangers of disrupting energy flows was not surprising.

The year 1992 was also when the world's diplomats converged on Rio de Janeiro for a historic UN Conference focused on Environment and Development (UNCED), also referred to as the Earth Summit. It wasn't the first time the UN had met to discuss human socioeconomic development and environmental challenges—indeed, it was held in part to commemorate the twentieth anniversary of the first Human Environment Conference. Rio's goal was to begin to design an approach to international cooperation to deal with the broad issues of how to manage mankind's burgeoning footprints on the planet, including climate change.[4]

Some will argue that it took much too long to get to Rio. After all, in 1856, American amateur scientist Eunice Foote made some of the first measurements that pointed toward the risks of a warmer world as a result of increasing carbon dioxide. Her long-overlooked work focused on adding carbon dioxide and moisture into a tube and examining thermometers placed inside. She detected a higher temperature when the tube was in sunlight. But Foote was clever to also put the tube into a darkly shaded place, suggesting that she understood the potential importance of infrared light. The twenty-first-century expansion of consciousness around women's scientific contributions has thrust her findings into the public eye after more than 150 years of obscurity. A few years after Foote, a British physicist named John Tyndall made similar measurements and received all the credit that should have been split with her. Whether he knew of her work is unknowable. What is evident is that Tyndall was far better known than Foote, having done important work on subjects such as the structure and motion of glaciers, so he was in a better position to make his work known to others. Tyndall also achieved a sharp distinction between visible and infrared radiation by using hot water to expose his tube to infrared energy (rather than the Sun), making his experiments a somewhat clearer analog of the power of carbon dioxide to potentially warm the planet.[5] But although, by the mid–twentieth century, researchers had firmly established the rise in carbon dioxide from fossil fuel burning, any associated temperature and climatic changes remained largely impersonal, imperceptible to the public, and subtle.

I once met a conservative congressman who challenged whether scientists could be sure that fossil fuel burning was really the cause of carbon dioxide increases in the twentieth century. "How do we know it isn't just coming out of the ocean?" he protested. I told him that we can use carbon dating. The same basic method of carbon isotope measurements that scientists use to estimate the age of a caveman's ice ax shows how the age of carbon in our air is changing. Along with their visible light, the sun and other stars emit high-energy particles. These particles bombard our atmosphere, constantly turning some atmospheric nitrogen into isotopic carbon, carbon-14. Living plants and animals, including us, use that isotopic carbon. The amount of this radioactive material decays by half about every five millennia (meaning it has a half-life of five thousand years), so, when we find a bone spear point that ancient humans made, we can measure the amount of this isotope to figure out approximately when that bone stored its carbon. The carbon in fossil fuels is very old, having been buried in the Earth millions of years ago. So as we burn those fuels and put their carbon into the air, atmospheric carbon dioxide gets "older." And it's doing that at just the rate we expect from global fuel use. We measure the age of carbon coming out of the ocean, too, and it's easy to show that it can't be coming from there. I remember watching the congressman nod reluctantly as he realized that this made sense. This interaction, like many others, showed me that members of Congress are not often stupid, although they are all extremely talented at acting like they are when it suits their political purposes.

Preparations for Rio energized the world's environmental diplomats. Many modeled their policy ideas and optimism on the incredible high the same people had just achieved in the arena of stratospheric ozone depletion. In the lightning-fast interval of five years, they had crafted a broad Vienna Convention in 1985 to consider ozone-depleting substances; then a binding international agreement in 1987, the Montreal Protocol; and by 1990 the production of the key ozone-destroying chemicals had fallen by an astonishing 50%. Why should the two issues not follow similar trajectories? Both the ozone-depleting gases and carbon dioxide have very long atmospheric lifetimes, and by 1992 the world had learned quite well the painful lesson first taught by Rachel Carson—that persistent chemicals often pose grave dangers to a safe and healthy Earth.

When the first steps to protect the ozone layer began, it was a "future" problem—a small effect not yet observed but viewed as a looming danger for the world in the twenty-first century—and climate change was the same in 1992. Through the process of dealing with ozone depletion, the science, policy, and technology communities had learned the power of assessments of the state of knowledge by broad teams of international experts. For climate change, the United Nations Environment Programme created an official Intergovernmental Panel on Climate Change (IPCC) to bring the scientific assessment tool into the climate process, and the IPCC had already delivered a first comprehensive report. What could possibly go wrong?

But while the diplomats expressed optimism, every one of them knew that the road to Rio would be rocky, uphill, and unique. They'd be lucky to get any agreement at all, not because of the science, but because of the deep divisions they'd have to cross—the vast cultural chasms that divided the global south, global north, east and west, and still do.

Europeans led the effort to work toward emission reductions at Rio, and it's easy to look at their stance naively and think they must just be better educated than we Americans, or more concerned about the environment. That's what I thought when I was a young scientist; however, as I spent time bridging the science/policy divide, I learned better. British Prime Minister and known arch-conservative Margaret Thatcher was one of the first to proclaim global warming a grave threat to humanity. In a speech at the 1990 World Climate Conference in the runup to Rio, she declared, "The danger of global warming is as yet unseen, but real enough for us to make changes and sacrifices, so that we do not live at the expense of future generations."[6] Some scientists nodded approvingly, noting that she held a degree in physical chemistry, so she had the background to understand the problem well. I'm a physical chemist myself. But I don't buy that explanation. I think that behind her statements was a labyrinth of economic and political incentives. The United Kingdom began to exploit North Sea oil and gas in the early 1980s. Replacing the traditional British coal with oil and gas was a useful gain for the environment, because oil and gas produce less carbon dioxide than coal to generate energy, as well as less air pollution. More important, British coal mining had also become unprofitable, and turning to the North Sea resource was great for the British economy. It also helped Thatcher break the back of the powerful British coal miners' union, and if there was one thing the free-marketeer "iron lady" hated, it was a strong workers' union. So conservative politicians promoted the transition away from coal.[7] But the British also knew they'd run out of North Sea oil and gas eventually, and like almost all other major European nations, they had a lot more to gain in the longer run from international pressure to reduce fossil fuel use entirely.

Nearly all major European nations are importers of fossil fuels. They have limited domestic sources of fossil fuels, so they are forced to buy them from others. Unrestrained fossil fuel use and sales are fantastic for the economies of countries that have enough coal, gas, and oil to meet their own needs and can sell the rest. That unrestrained use also puts those who have to buy it from others at a huge economic disadvantage. Better and cheaper non-fossil resources will become available if international regulation sparks new technologies and increased supply of those technologies. That will be a huge benefit for energy-importing nations.[8] While the Europeans might seem like angels in this problem at first glance, the old adage to "follow the money" usually helps to demystify environmental negotiations. Reducing energy dependence on foreign sources also strengthens security against political events, instability, or whims in other nations—as exemplified by the oil shocks of the 1970s associated with turmoil in the Middle East, and later by the Russian invasion of Ukraine and resulting war in 2022.

The United States has vast energy resources. We had long been the Saudi Arabia of coal when the Rio meeting began in 1992, producing more at that time than any other nation, with hefty oil and gas industries as well. Our economy had boomed using the engine of our rich fossil fuel resources for decades. An agreement to curb their use was not in our economic interest. And the drumbeat of "It's much too expensive to do anything" had already gained traction in a nation that had just experienced the presidency of Ronald Reagan. We were opposed to any agreement at Rio, and our position has fluctuated over the years depending on economic tides and political aspirations.[9] While China now holds the title of world's largest coal producer, today we produce more oil than Saudi Arabia, due to changes in drilling technologies.[10] The difference between US leadership in international efforts to combat ozone depletion just a few years earlier and our on-again, off-again approach to climate change at Rio and beyond is striking.

Within the lower-income parts of the world, the foremost priority is poverty alleviation and development, at the fastest possible speed and therefore with the cheapest possible resources. Just one trip to one of the world's poorest countries is guaranteed to jolt even the most hardened resident of a wealthy nation into understanding the essential rightness of that imperative. It takes energy to build roads, electrify a nation, produce food and clean fresh water, install indoor plumbing, and build more hos-

America's show windows have plenty to show

— thanks to COAL!

It's difficult to name anything Americans build, buy or use that doesn't take coal. Think of the steel that goes into your dishwasher, your son's bike, your auto—every ton of steel that's made takes the carbon from a ton of coal. Think of the electricity that runs your appliances, lights your lamps—70% of the fuel used to produce it is coal. And thousands upon thousands of the fine products that make our American standard of living the highest in the world are manufactured with power generated from coal!

How fortunate, then, America's coal resources are large enough to furnish all the power, light and heat this country can use for centuries—that America's coal industry is the most mechanized and efficient of the world!

Are you responsible for choosing a fuel to power a factory—to heat a home, apartment house or other building? Then you should consider these important

ADVANTAGES OF BITUMINOUS COAL!

☆ Lowest-priced fuel almost everywhere!
☆ Labor costs are cut with modern boilers and automatic handling equipment!
☆ Easiest and safest to store of *all fuels!*
☆ Vast reserves make coal's supply dependable!
☆ Dependable supply assures price stability!
☆ A progressive industry strives to deliver an ever better product at the lowest possible price!

BITUMINOUS COAL INSTITUTE
A Department of National Coal Association, Washington, D. C.

FOR ECONOMY AND DEPENDABILITY
YOU CAN COUNT ON COAL!

Time magazine advertisement by the Bituminous Coal Institute, 1952.

pitals. The innumerable blessings that people in the rich parts of the world take for granted are desperately lacking in many places. It is understandable that the lower-income world has very little interest in agreeing to reduce emissions if it would hinder their attainment of a healthier and more comfortable lifestyle.

Political distinctions can be found within lower-income nations, too. On the one hand, there are the oil-rich countries like Venezuela and the Gulf states, whose positions at Rio stemmed from some of the same origins as those of the United States. About 20% of annual carbon dioxide emissions currently come from deforestation, and countries like Brazil and Indonesia take a dim view of any controls that might threaten their sovereign rights to use their own forests as they see fit. And a lot of lower-income nations felt that attempts to institute a global UN treaty on climate change was just another way to impede their progress, a new and sneakier tool for oppression—a fresh form of colonialism. On the other extreme is the Alliance of Small Island States—places like Samoa, the Maldives, and the Bahamas, where there is every reason to fear that climate change and accompanying sea level rise will literally drive them off their tiny patches of land within the twenty-first century.[11]

Those of us lucky enough to live in the richer parts of the world (sometimes called the global north, which includes the United States, Europe, and perhaps paradoxically, Australia and New Zealand) are currently emitting far more carbon per person than those in the poorer global south—by a staggering amount. For example, in 2021, on average, an American emits about a hundred times more carbon than the average Ethiopian, thirty times more than the average Nepali, twenty times more than the average Nicaraguan, and twice as much as the average Chinese.

Fossil fuel emissions are increasing and are expected to increase sharply in this century because the countries of the global

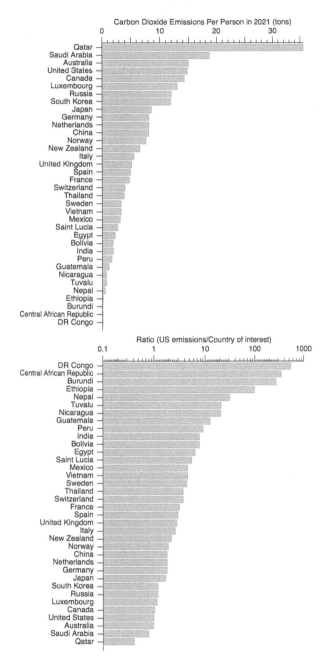

Carbon dioxide emissions per person (metric tons per year) in 2021 for se-
lected countries (top). Ratio of US per person emissions in that year to those
countries (bottom).

south are finally beginning to attain more wealth, to lift their people out of poverty by development, as China has done. But if ongoing twenty-first-century global development is powered by fossil fuels, as ours was and largely remains, then the planet is in for a warming of about 3–5°C by 2100 (a "business as usual" scenario), and that's dangerously hot. High average temperatures are just the beginning of the havoc due to the heat waves, sea level rise, crop losses, water shortages, wildfires, and other impacts that would surely be associated with such levels of climate change. And even more warming would follow. The high per capita fossil fuel use by rich countries also needs to decrease dramatically if global warming is to be curbed.

Research indicates that several already hot and humid regions (for example, in coastal parts of the Middle East) would be the first to become virtually uninhabitable, because the human body just can't cool itself enough (by sweating) beyond a certain temperature in humid conditions.[12] Some people argue that this won't happen because we will run out of fuel before there would be any problem. That's just plain wrong. For example, researchers have estimated that burning all the possible coal resources on Earth would make the average temperature on the planet a scorching and unlivable 14°C, or about 25°F hotter.[13] Some scientists dispute how accurately we can make such projections, given the complexities of the climate system and uncertainties in just how much coal can be mined, but smaller amounts of warming such as 11°C or 12°C would make not only the Middle East but the current homes of most of the world's population uninhabitable.[14] Catastrophic droughts, crop failures, and water shortages would also be unavoidable.

As terrifying as these extreme cases are, we are also seeing plenty of damage at current levels of warming, and every tenth of a degree of further warming also means human suffering and death, and damage to nature. Looking at some of the examples where we have easily available data, researchers estimate that every tenth of a degree of warming between 1.5°C and 2°C will lead to about a 5% decrease in water availability in the Mediterranean, 2% more global corals at risk, and 2% less wheat to feed the world. These are frightening numbers in a fragile world.[15] There is good reason to be concerned about similar but unknown effects in other places and systems. Banking on uncertainties in future climate projections becomes foolish in my view. The models would have to be drastically off to reduce the risks to levels that would seem safe and comfortable for the world if fossil fuel use continues to be our main source of global energy. And the uncertainties cut both ways: things could just as easily turn out to be worse than the models now predict, as was true for the unexpected Antarctic ozone hole.

The island nations emerged as the foremost group pushing for controls in Rio. Others weighed the potential for an agreement they might find acceptable if (and only if) the richer nations went first and agreed to pay the cost of poorer countries' development with clean energy. After all, most of the carbon dioxide and climate change so far came from the emissions of the rich, so having created the problem surely it was their responsibility to start fixing it first. On top of that, the rich can bear some costs of adapting to climate change rather than trying to reduce it, but the global south just doesn't have the resources needed to insulate themselves from the ravages of a changing climate.

For a scientist like me, an amazing part of attending any international negotiation is how progress feels like watching paint dry. The diplomats meet, and meet again, and talk around each other, and try to stall or manipulate in a way that drives any normal person to distraction. How incredibly inefficient they are, I thought at my first UN meeting. Why don't they just get to

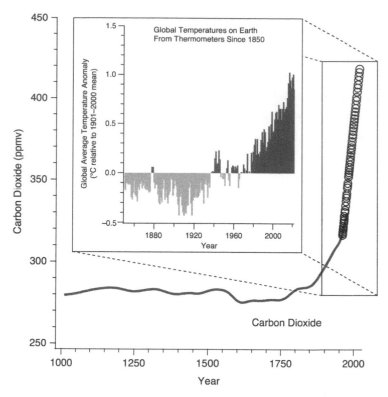

Carbon dioxide concentrations from ice cores and other records along with modern measurements. Inset shows global averaged land and ocean temperature anomalies, relative to the average over the full period.

the point? And the climate negotiations take inefficiency to new heights (or should I say lows?). At any UN meeting, the first order of business is agreeing on the agenda, a task one might think was easy enough. The country representatives were seated in rows, with their nation's name on a plaque in front of each group of a few people. We were in an imposing hall of considerable size, and over a hundred nations were represented. Scientists and others from authorized groups (e.g., the IPCC, Greenpeace, etc.) sat at the back. But as the diplomats discussed it, the hands of the clock marched right through the proposed time for a coffee break, then past the lunch break. There was still no agreement even on when these would take place, so by default they didn't.

It wasn't because of lack of understanding. They understood each other only too well because we all had our headphones on so that every word could be simultaneously translated into the six official UN languages: English, French, Spanish, Chinese, Arabic, and Russian. At first, I chuckled to myself. As hours passed, I realized that the real subject matter of the meeting was just so supercharged, so emotional and vital, that the painful debate on the agenda was the diplomat's way to test their opponents and the chairman up there on the grand stage, throw others off balance, and stake out their country's dominance in the negotiating process. Only after enough huffing and puffing had taken place could they approve the agenda and begin the meeting. The subject that was on the table that week was the question of whether and how to account for proposed carbon sinks such as reforestation. For example, how can policies ensure that commitment to reforestation in one location doesn't get canceled by devastating an existing forest in another spot? Do countries have to agree to outsiders coming in like policemen to monitor and measure, and how respectful of local sovereignty is that? These questions are not easy to address, and it took several years of repeated meetings and special scientific reports to make very limited progress. Diplomatic meetings take place in extreme slow motion for any scientist, but you do learn a lot about the seemingly alien world of international environmental diplomacy.

The nations of the world did make a broad agreement on climate change at Rio, called the UN Framework Convention on Climate Change (UNFCCC). Not surprisingly, given the disparate viewpoints of different countries, it did little more than establish grand objectives and principles and the hope that nations would work to reduce emissions to meet them. It did achieve useful agreement to require countries to report their national emissions, but did nothing to reduce them. Substantive future decision-making would depend on many things, especially the evolution of value judgments about how much risk is too much in disparate cultures with wildly varying priorities. The Paris Agreement is a part of this broader convention today.

Some people love skydiving, but I'd never do it—I just don't have the same feelings about that risk, and it's a personal choice. But our national and international climate change choices force involuntary risks onto others. Although scientific information can help to build clarity and consensus around those risks, our collective values are deeply embedded in our climate change choices.

In the early 1990s, a Japanese energy economist named Yoichi Kaya came up with a clever way to think about the things that influence the emission of carbon dioxide from burning fossil fuel and industrial sources, the Kaya identity. In my opinion, it's a stunning, simple way for everyone to absorb the enormity of the task of taming climate change. The simple little equation works both for individual nations and globally. It states that emissions (call them F) come from multiplying four factors together: population (P), gross domestic product (G) per capita (how rich we are per person), energy consumption per unit GDP (E, how much energy it takes to make a unit of money), and how clean that energy source is, or fossil fuel emissions per unit energy. It's that simple: $F = P \times (G/P) \times (E/G) \times (F/E)$.[16]

An exasperated student once asked me why I think this equation is essential for anyone trying to understand climate change. It's an identity, he protested, so what good is it? In mathematical terms, an identity holds for every value of its terms—it just has to be true by definition. That's a fact, I replied—but it doesn't

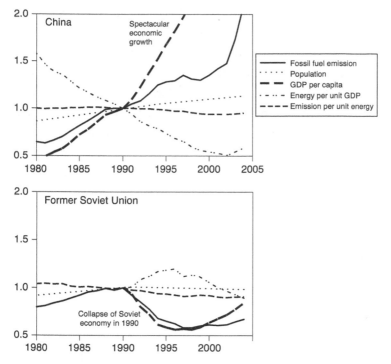

Factors in the Kaya identity $F = P \times (G/P) \times (E/G) \times (F/E)$, where F is carbon dioxide emission, P is population, G/P is GDP per capita, E/G is units of energy used per unit GDP, and F/E is carbon dioxide emission per unit energy, for China (top) and the former Soviet Union (bottom). All factors are normalized relative to 1 for 1990.

mean the equation isn't useful. And the test of its usefulness is whether it helps us better understand the factors driving emission changes over time in various countries.

No examples are better than looking at the past emissions in the former Soviet Union and China in the Kaya framework. The population of the Soviet Union was nearly flat from 1980 to 2005, and its energy use per unit GDP as well as fossil fuel emission per unit energy hardly changed either. But after 1990, its emissions took an abrupt nose dive. That's because the collapse of their Union was accompanied by huge economic problems and a crash of its formerly strong GDP. In contrast, China has managed to decrease the energy it needs to make a unit of GDP, and the fossil fuel emissions per unit of energy have also declined over that time. This shows that its economy is actually getting cleaner. Its population increased a little over this period, but not much. Its fossil fuel emissions have skyrocketed mainly because it has experienced explosive economic growth, especially after 2002 or so. It's not lost on my students that this comparison has a crystal-clear conclusion: if China, or any other nation, wants to substantially cut its fossil fuel emissions, then it would have to either curtail its economic growth or greatly accelerate its use of clean energy. The Kaya identity lifts the veil on understanding why international climate change negotiations often feel like a Kabuki dance and are so horribly fraught.

Questions of costs and benefits are sharpened in this environmental issue as in no other. But a balanced weighing of benefits versus costs has proven elusive as well as controversial in scholarly circles. Benefits are difficult to quantify and involve many intangibles and questions of values, such as how much is a human life worth? A life is invaluable if it's yours or your family's, but for the purposes of cost-benefit analyses, the EPA and others design technical methods for valuation that put the value of today's average young American at about $10 million, which is designed to reflect a lifetime of average earnings and contributions to the national economy. The old are worth much less in this rubric—but not to themselves. And the value of species that may go extinct, or of ecosystems that may be damaged involve personal values and not simply numbers. So the benefits of nature are often simply left out because they are so difficult to monetize. In contrast, the cost of clean energy can usually be estimated, and no nation wants to impede its GDP in any way. Thinking only of known costs and not of benefits skews the analysis in a way that sells life and the environment short—and opens the door to fear, anger, and international conflict.

I'm often asked whether climate change is a population problem. That's not the main issue at this point, I respond, although such remarks often draw boos from the ardent protectors of the planet. I'm an ardent protector too, but I have to look at the numbers. You could argue that the planet has far too many people for the energy system we use—fair enough. But that system is already here, and so are we. Almost 8 billion of us were on the planet in 2020. Further population growth is expected to reach about 10–11 billion within this century, which will worsen the situation by some 20%–30%. But as in the case of China's explosive change in emissions recently, it's global development of poor countries that will drive the great bulk of twenty-first-century carbon dioxide emissions (unless it occurs with clean energy). At present, the roughly 6.5 billion people in the global south emit on average about a quarter to a third of the carbon per person than the other 1.5 billion who live in the global north (and in the rich communities of certain countries such as China and India). So if the world's 6.5 billion impoverished people use fossil fuels to get rich, as the United States already did, then emissions will soar—doubling or even tripling. Another 20%–30% in total population makes it worse, but the fundamental problem is that it is only the tragedy of global poverty that has kept us from burning to a crisp already. You have to be inhuman to be against development and the associated reduction in human suffering and privation, but twenty-first-century societal development has to be powered another way to avoid massive climate change.

Safeguarding our planet's climate therefore depends on powering the entire planet, from south to north and in between, with much cleaner energy, and doing it without breaking the bank. We are very wasteful with energy today, so of course conservation is helpful. Reducing our emissions of greenhouse gases other than carbon dioxide, like nitrous oxide, and methane, as well as soot, and other pollutants also helps mitigate climate change. Agriculture and food are sources of several of these emissions. A diet of meat rather than plants is worse for climate change, partly because raising cows and sheep requires more energy use per food calorie than plants. All of these, and more, have been cited as factors in solving the climate change challenge. They do merit consideration, but they are not the key priority in my view. Carbon dioxide is the primary factor causing our climate to change today, and if you accept the premise that the world is not going to go live in caves and wear animal skins then some level of modern energy use is fundamental. If the global south gets its energy for development from fossil fuels and we in the global north continue our outsized consumption of fossil energy, then carbon dioxide forcing of changes in our climate will only become ever more dominant.

Controlling the ozone-depleting substances under the Montreal Protocol was relatively easy for many reasons, not least of which is that these gases only represented a small fraction of the total business for typical chemical companies in 1987 (2% at the DuPont company, for example), and the economic cost of replacing that technology nationwide in the United States in 1990 was estimated at only about three billion dollars[17]—a very small fraction of national GDP. But the GDPs of nations worldwide are tightly linked to energy production, as Kaya brings into sharp relief. And the estimated value of the modern global fossil fuel industry towers over other environmental challenges, at over 50 trillion dollars.[18] It's a staggering figure. Every coal company owns valuable mineral rights to mine coal, and large investments in expensive machinery, transportation resources, and much more. Every oil company has mineral rights too, and among other investments perhaps dozens of offshore oil platforms, each of which costs at least 200 million dollars. These assets risk being "stranded" if their products can no longer be mined and sold, providing a colossal incentive to make sure that climate change is not a priority for action anytime soon.

After dragging their feet for a few years, the chemical industry pivoted to making replacement compounds for the ozone-depleting gases fairly quickly after the Montreal Protocol entered into force. In contrast, it is not likely to be the fossil fuel companies who will make solar panels and wind turbines, so nearly all of them have long resisted new energy policies with all their might (and their clout and economic resources are indeed mighty). Every day of delay in the world's actions on climate change is a bit of those fossil fuel company assets used rather than stranded, so their business goal will logically be to delay as long as possible. It is to their tremendous credit when these companies realize the days of dirty fossil fuel use are numbered, and begin genuine efforts to make their businesses cleaner. Some of them have done that by increasing their emphasis on research to capture carbon emissions at power plants, for example, while others have tightened up methane gas leaks. These actions are a start, albeit a small one.

It was so easy for citizens to take personal action to start reducing ozone-depleting gases—all we had to do was switch from spray cans to roll-on deodorants to kickstart the ozone loss problem onto the path toward solution. With three-quarters of global use of chlorofluorocarbons in spray cans in 1974 (a lot of it in those "personal care" products), this simple act packed a wallop. But meaningful steps to address climate change require much more than personal choices. While riding a bike, changing your diet, or switching out your light bulbs helps a little, such acts are very far from a global climate change solution. The scale of the necessary change in global carbon dioxide emissions requires changing the world's entire energy system, something that the individual consumer just cannot do on their own.

Does this mean changing this gargantuan industry is impossible? Surely not. After all, we achieved it when the powerful auto industry was made to clean up its act (and along with it American smog), and when the sprawling farming and pesticide industry was forced to stop its love affair with DDT. Just as in those issues, our most important choices are those that generate political pressure and new policy decisions—the things we do as citizens, voters, and yes, peaceful demonstrators—and groups have more power than individuals.

There are myriad barriers to social action. In the 1960s and 1970s, centrists and left-leaning individuals largely led environmental progress in the United States, but had substantial bipartisan support from some conservatives. But beginning with the Reagan regime's attacks on regulation and the EPA, conservatives became less interested in conservation and more interested in attacking environmental protection. They quickly learned that the full-frontal assault on government action was less effective with the American public than guerrilla warfare against the underlying science. As the power and influence of science grew in environmental action and decision-making, it naturally became a prime target for those interested in inaction. Well-funded by wealthy families such as the Koch brothers, the fossil fuel corporations, along with conservative foundations, activists, and groups, began an organized campaign of disinformation, inventing new buzzwords like the need for "sound science" as opposed to their view of climate science as unsound, or even better, "junk science." The rise of this line of attack dovetailed neatly with the growing use of the Internet and the spread of climate disinformation.

When I write a scientific paper, I submit it to a journal, where the editor picks out two or three expert peer reviewers who have experience with the issue at hand, and asks them to give it a careful and critical examination. As a reviewer, I study papers I get for review very carefully, and spend many hours figuring out whether I think the analysis is complete and correct, and then I write reviews that include extensive questions and issues for the authors to respond to. They have to write a point-by-point response, and the paper is generally changed, then re-reviewed to make sure all of the peer reviewers' comments are adequately addressed. It's not unusual for review, response, re-review, and re-response to occur. This book was also stress-tested by peer review (although, of course, all remaining errors are my own). Rigorous peer review and publication is a slow process, with scholarly work typically taking months or more to appear after all that, and it weeds out a lot of incorrect or misleading publications. Even when one slips through this careful scrutiny, science is intrinsically self-correcting, because ongoing work is constantly testing previously published findings, and any major mistakes are typically caught within a few months or so. In contrast, anyone can post their unreviewed, untested, and far-fetched claims on the Internet for broad consumption at the touch of a button.

The Internet, along with the ferocious opposition of the fossil fuel industry and the advent of conservative think tanks led to attacks against climate change that were far more potent than the tamer ones that occurred in the heyday of environmental policy in the 1970s. These had a resounding effect on US public opinion and political action in the 1990s and 2000s. While this movement began in the United States, it has spread to Europe and beyond. Sometimes dubbed the culture wars, a broader social movement has also had a profound influence, leading a range of different actors and sectors of the citizenry to latch onto these ideas and oppose environmental protection and particularly action on climate change—especially people who feel resentful of academics or left behind by social trends. Today we have lived through a remarkable swing of the pendulum, away from the social movement of the 1960s and 1970s that brought us cleaner air and water and saved our ozone layer, to a new social movement that has created deep polarization around the environment along with most everything else.

Climate change deniers have used a range of methods to attack climate science and individual climate scientists, including for example the hacking and misrepresentation of emails sometimes referred to as "Climategate." This involved a hack of the mail server at the University of East Anglia in England, providing fodder for distortion and accusations against scientists participating in the IPCC of politically altering its reports. Another approach to attack the science is sometimes called the "weaponization of uncertainty." Spreading doubt by raising the question of how uncertain climate change data or models are, or how accurate future projections of such a complex system can be, is an easy target for attackers. The approach is textbook. It was used, over decades, for example, by the tobacco companies to delay admitting to the health damage and deaths from smoking, allowing them to keep selling cigarettes as long as possible with no warnings.[19] There will always be some scientific uncertainty in any subject—science continually strives for an exactitude that is impossible to attain. It's a convenient target to attack and to exaggerate in order to camouflage the power of science: the well-understood physics, and the grave risks of climate change.

Climate change disinformation, especially that produced by think tanks and the scientists they support, often takes the form of true but irrelevant claims. A good example is the argument that "the ice ages weren't caused by people so how do we know we are the cause of climate change today?" At the peak of the last ice age about 20,000 years ago, the average temperature on Earth was roughly a chilling 5°C colder than today. We know that Earth emerged from that into the milder climate we enjoy today about 11,000 years ago, at a time when our ancestors had virtually no carbon footprint. So what happened?

First off, I like to tell my students that the whole planet can't have climate mood swings. Earth requires a gain of energy to warm overall, or a loss of energy to cool. Local changes are another matter, because energy can easily flow from one place to another on this planet, but on a global basis nothing can magically make energy or destroy it. The ice age, like today's climate change, was a response to shifts in forcing, and it also was linked to feedbacks that reinforced those shifts. Second, the fundamental reason that the ice age came and went is that Earth is like a spinning top, or a gyroscope, that undergoes big changes in its orbit on extremely slow time scales of many tens of thousands of years. These dwarf the small changes over seasons and decades that the orbit goes through now. The distance between Earth and the Sun changes, as well as its angle relative to the Sun, and these drive massive changes in climate. The changing distance from the Sun brings us more or less solar energy, and the angle alters the amount of light hitting the poles. When the poles receive less sunlight, they get a lot colder, and ice can then accumulate and grow; we can see this in miniature each year as snowpack and sea ice builds up in winter in many places. But unlike these annual ice cycles, the ice accumulation of the ice age didn't get interrupted by a warm summer because the Earth's orbit was so different then that Earth just never got warm enough for the ice to melt at any time of year. The poles became ever more reflective as a feedback, bouncing even more of any light they did get out to space. That made high latitudes get even colder, and even more ice grew, in a vicious cooling cycle year after year. During the last ice age, Chicago was under the north polar ice cap, and the massive front of ice spreading over North America carved out our Great Lakes. Ironically, rather than making you feel complacent about global warming, this past climate change should give you more reason for concern. It wasn't a mood swing. It came and went because of changes in planetary energy caused by Earth's multi-millennial orbital changes, but it is we humans who are much more quickly changing our planet's energy flows now. We know that the Sun's radiation, measured by satellites, has not changed in intensity over the past fifty years. So where did that extra energy come from to warm the planet up by about a degree Celsius since 1850 as observed? We know that the necessary energy to heat the atmosphere as observed isn't somehow coming out of the world's oceans, because they are warming up along with the atmosphere. There is no magical source for that energy, but there is indeed a scientifically well-documented one: the change happening in the energy budget comes from our human-caused emissions of greenhouse gases, especially carbon dioxide.

Another example is the weaponization of the statement that water vapor is the Earth's dominant greenhouse gas, not carbon dioxide. It's true, and it's also irrelevant. Indeed, water vapor does absorb infrared radiation strongly, but it responds to the climate instead of driving climate. Water vapor has a very short lifetime in the atmosphere—typically days. The water in the atmosphere is controlled by the balance between evaporation at the surface (from the oceans, but also from soils or plants) versus removal by formation of clouds and rain. Evaporation in turn depends upon surface temperature. Consider for example, the humid air of a tropical jungle versus the extreme dryness of Antarctica, even coastal Antarctica. Changes in humidity are a response to climate and not a forcing of it—it's not going to make the climate change, it's going to react to a change in climate.

Carbon dioxide can force the climate into an altered state precisely because of that pesky persistence problem in environmental issues. Just like DDT, lead, and the chlorofluorocarbons, this looming problem reflects many years of human activities piled up on top of each other. Carbon dioxide's lifetime is so very long in the atmosphere that it builds up year after year as we go on emitting it by burning fossil fuels, and the climate has to respond. While some carbon dioxide we humans emit is removed on a time scale of a decade or so to forests and plants, that loss quickly reaches its limit because plants can only grow so fast. Some is taken up by the ocean on timescales of decades to centuries, but that also reaches a limit ruled by chemistry in the upper ocean (the same chemistry that means that increasing carbon dioxide causes ocean acidification). And here's the kicker: even in 1000 years, some 15%–20% of the carbon dioxide emitted today will still be in the atmosphere, continuing to warm the planet and acidify the oceans. This has been a subject of some of my own research, and it shocked me more than any other aspect of science that I've ever worked on. It also made me reflect deeply on my own choices. Gulp, I thought, when I first realized that even my unnecessary trip to the grocery store to satisfy a craving had put carbon into the air that would last for a thousand years. More important, like others in the rich world, I know that I'm responsible for many tons of long-lasting carbon per year even though I try to cut it down. Some people imagine that we will come up with some clever and cost-competitive way to suck carbon directly out of the atmosphere and put it away where it won't come back. The current costs and enormous amounts of the carbon that would need to be removed limit such approaches, but they can contribute and most scientists expect that they will—at a modest level. But we don't have such technology at workable costs and scales now, so it would be extremely damaging to rely on that magical future invention instead of ramping down our carbon dioxide emissions.

Another techno-fix for the climate change challenge is to devise ways to block the Sun, harkening back to the way a major volcanic eruption can cool the planet for a while. This is sometimes called *geoengineering*, but I think this is a misleading term. We use the word *engineering* when we understand the principles and know just how to do something, like building a bridge. We can even build models of the bridge and test them to be confident they're going to be very safe before we even build them, and then we can test the built structure before we let people cross them. And while we might make very rare mistakes, we can calculate tolerances and safety. But not for geoengineering. I think of the idea of climate fixing not as geoengineering but as climate tinkering or worse. It is feasible to load the atmosphere with haze to reflect energy out to space and cool. That's what happens after a big volcanic eruption for a few years, so yes, it can cause cooling. And it is surely cheaper than decarbonizing our global energy supply, but it also has a host of dreadful consequences and risks. One is that carbon dioxide would still build up, and cause ocean acidification. While scientists understand the chemistry well, they do not equally understand the biological effects of a more acidic ocean. Given evidence of five mass extinctions in the Earth's past five hundred million years—all associated with carbon and the ocean—taking on the risk of triggering a sixth in this way seems foolish. Another is the strong likelihood that action to block energy from the Sun would affect rainfall. Imagine the global strife and havoc that would ensue if climate tinkering by rich countries to avoid cutting our emissions caused a destructive drought in even a single poor one. The toughest issue in international negotiations is the loss and damages from climate change experienced already in poorer countries as a result of emissions by richer ones—even though today's climate change isn't something the wealthy intentionally inflicted on them. Imagine what would happen if we did things on purpose in the future that ended up making just one (and maybe more) other country's climate worse. And what if our deliberate climate tinkering led to ocean acidification that damaged the corals or shellfish some nations depend upon.

The potential for disastrous side effects brings up the question of whether any plan to tinker with the climate on a global scale to avoid mitigating our impacts would be governable.[20] Many international relations scholars have argued that reaching global agreement on climate tinkering would be impossible,[21] in part because all countries would be affected by what is done and some (probably the poorest) would lack the ability to adapt, perhaps disastrously so.[22] It's virtually certain that only a limited number of countries would want to proceed with it, and how could that occur without conflict? Making matters worse, they'd all have to cooperate and abide by any agreement rigorously for several decades at least, something the world hasn't managed to achieve with what seems to me to be the simpler issue of nuclear non-proliferation. In my view, we are faced with a dilemma of what I call "creeping normalization" around geoengineering the climate if research and discussions proceed without governance issues being made crystal clear first.

Capturing carbon before it even gets released from a smoke-stack is a very different approach sometimes also called geo-engineering, even though it's very different. It is already done to some extent at power plants and large industrial plants. The captured carbon dioxide is sometimes stored by pumping it back down into the ground (e.g., in depleted oil wells) where it can remain. So far so good. How much comes back to the atmosphere or leaks is one concerning issue. And while this technology has potential, if the carbon dioxide is used instead of stored (for example by pumping it into a working oil well to flush out more oil), then any gains for the planet can become murkier. Perhaps someday carbon dioxide can be used economically but not re-emitted: for example, to make products like carbon fibers. But such practices won't become practical until they are profitable, and for now carbon dioxide capture and subsequent use remains largely a hope for the future.

The ozone-depletion issue became a hot crisis in the late 1980s, after scientists discovered the Antarctic ozone hole. The discovery was dramatic and easily perceptible in the imagery from satellites, which reveals how ozone is rapidly gobbled up to form a gaping hole over the entire continent in just a few weeks as sunlight returns to the cold polar cap every spring. Scientists rushed to investigate, and scenes of us working in the coldest place on Earth could not fail to capture public interest. In contrast, climate change remained stubbornly slow during the twentieth century, so that public understanding and perception of the problem were limited, despite the unstinting efforts of many a great scientific communicator. The social trajectories of hot crises command public interest, political attention, and global action, as demonstrated across the spectrum of challenges from the ozone hole to Ebola, while issues like climate change and endemic hunger fester on the back burner, outpaced by whatever other urgent global matter emerges next.[23]

Sometimes I've mused that it would take a huge chunk of Greenland falling suddenly into the sea to wake the whole world up to the crashing sound of rising sea levels and make climate change into a hot crisis, but I know that the best scientific work doesn't suggest that will happen. Instead Greenland and Antarctica will gradually shed ice on slow time scales—centuries and more—allowing humankind to adjust constantly to altered new climate normals of sea level. How about other potential hot climate crises? Climate change involves phenomena like wildfires, floods, and hurricanes, that, although terrifying, are familiar. We've seen them before, so to understand climate change means understanding that they are happening more often, or more severely. This means it comes down to statistical concepts that are not easy to grasp at first. In contrast, the threat of severe ozone depletion carried with it the threat of something never before known to humans: bombardment with dangerous rays of ultraviolet light that can hurt all of us. While lighter-skinned people more often suffer from skin cancer caused by excess ozone loss, eyes both brown and green become clouded with cataracts when too much ultraviolet light hits them in an ozone-depleted world. Damages to plants, animals, and ecosystems are less well studied but clearly very likely. On top of all that, ozone depletion made everyone but the chemical companies a loser; there were no winners. But it can be argued that the coldest countries, such as Russia or even Canada, might in principle benefit from a warmer climate, as long as they are willing to overlook resource wars, unmanageable immigration, and the tragic suffering of others.

As I became deeply involved in the issues of ozone depletion and climate change, I came to view the first as a stellar exemplar of how events sometimes line up just right for humanity to rise to the occasion on environmental issues, while the second repeatedly defeated us—until about the past five years, when many of the same factors that led to earlier environmental successes have come into an encouraging new alignment.

Today's changes in temperature, rainfall, and other aspects of the climate around us are occurring fast, with discernible changes roughly every decade. Even though our climates do vary, when changes get big enough that they push outside of past variations, we notice them. A great example is the increased breaking of hot temperature records. At more and more sites in remote areas with data going back at least to 1900, new high-temperature records are being set almost every year (and, importantly, few cold records).[24] It's fascinating (albeit sad) to study those changes worldwide. It's a great field for young scientists, not only because they can contribute to solving one of the most challenging environmental problems the world has ever faced, but also because they will spend their careers witnessing and probing how changes will emerge ever more strongly from the up and down of natural variability. Obvious and undeniable problems are already emerging in phenomena ranging from declines in particular fish populations as the seas get too warm for them, to the spread of horrific wildfires and devastating storms—there will be more to follow.

It is no longer just climate scientists who are aware of these changes. In Europe and the Andes, people have observed the retreat of mountain glaciers with their own eyes, and glaciers are retreating almost everywhere on the planet. The few glaciers that are advancing because of increases in snowfall (a favorite fact of some skeptics), are irrelevant to the global issue. Elderly New Englanders have told me that the high tide is coming in much higher on its rock-solid shorelines than it did when they

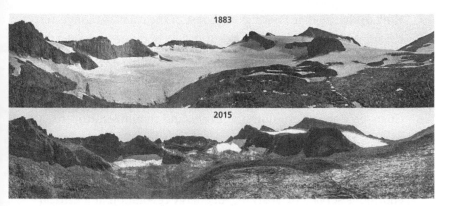

1883

2015

Photo pairs show the retreat of the Lyell Glacier (Yosemite National Park) between 1883 and 2015.

were children. Midwesterners and Europeans have noticed an increased frequency of the heaviest rainfall, and more rain is falling in extreme events like hurricanes and typhoons throughout the world. Tropical summers are hotter than they used to be throughout the Caribbean, Asia, Central America, and Africa. In many countries, it's become obvious that heat waves are more frequent, intense, and deadly. And with this perceptibility, climate change is also becoming more personal as people begin to fear for their own safety and that of their children in the face of a phenomenon they now experience themselves. The emergence of a personal and perceptible problem, that key step in widespread public engagement, is now well underway as the world warms.

While the climate has been slowly warming, technology and innovation was also advancing, spurring several countries to realize that whoever could develop the best cleaner energy methods could reap substantial profits. European leadership in clean energy use over many decades was not the lightning bolt for climate change that Ed Muskie's boost of the catalytic converter for smog reduction in the 1970s was, but both are examples of the power of policy, even "soft" policy, to drive technology-steering. Concerted European increases in their use of solar and wind energy began expanding demand and driving innovation. Costs go down as volume increases, and manufacturing rises to meet the challenge of increased production. The first unit of anything costs far more than the millionth one off the assembly line. So those who went first did the rest of the world an enormous service by sparking an energy revolution at a critical time. Sensing the market opportunities, China quickly became a world leader in the provision of those solar panels. US President Barack Obama put cleantech high on his list of personal priorities. He recognized the opportunity to make lemonade out of lemons when fate handed him the presidency on the heels of the 2008 financial crash. Experts point out that among the many moving parts of Obama's historic American Recovery and Reinvestment Act of 2009 (designed to stimulate the economy out of severe recession) was the biggest energy bill in US history at that time, topping out at around $90 billion dollars.[25] In a promising trend, it's now been dwarfed by the massive clean-energy stimulus of the Biden era Inflation Reduction and Infrastructure Act bills that followed in the 2020s.[26]

The French government embarked on a meticulous planning process in advance of hosting the key international negotiating meeting at which a new climate change policy would be finalized, the Paris Agreement of December 2015. Negotiation had taken years to recover from a disastrous unraveling at a meeting in Copenhagen in 2009. That meeting had been a shambles at which it became crystal clear that the old model of a small, male-dominated, and graying group of representatives of high-income countries agreeing among themselves and only later cajoling, bribing, or forcing the low-income nations to go along had to be consigned to the dustbin of history. It would be an enormous challenge to make Paris work differently, but the French team is widely acknowledged to have accomplished it with aplomb.[27]

At Paris, the negotiations formalized what history and the Kaya identity make obvious: while reining in climate change has much in common with other environmental agreements, it has progressed at a snail's pace because world society has taken a drawn-out approach to coming to grips with its unique challenges. The two countries that emit the most carbon dioxide are the world's current financial leaders, China and the United States. Neither will submit to a binding UN protocol to reduce their emissions on a set schedule, because that might interfere with their prosperity or sovereignty. Many other countries hold the same basic view. A more flexible arrangement had to be pursued if the deadlock that had produced so little international climate action, dating back at least to Rio in 1992, was finally to be overcome. Paris triumphantly came up with a new plan described as pledge-and-review, a system in which each nation submits an emission reduction goal, or pledge, and progress toward the goal is periodically and jointly reviewed UN-style, by representatives of all involved nations.

The central tenet of the Paris Agreement is to hold global average temperature increases to "well below 2°C above preindustrial levels" with an aspiration of "pursuing efforts to limit the temperature increase to 1.5°C above preindustrial levels."[28] Critical practical issues are still being worked on, particularly those surrounding transfers of funds from high-income to low-income countries to assist the latter in meeting their declared national goals. Similarly, the exact nature and specifics of the "loss and damage" provision intended to compensate poor nations that are particularly hard hit are still being formulated. The small island states are among those who ensured that these measures, which were sidestepped in Copenhagen, would take center stage in Paris. The Paris Agreement has been roundly criticized by those who believe that only a binding international agreement will be meaningful, but it is increasingly defended by many who see its pragmatism and the way it is driving practical solutions.

The Paris Agreement creates an impetus for nations to rise to the occasion, to be bold yet realistic rather than cautious. I sometimes like to joke that it's like Weight Watchers, a diet program that has succeeded for millions of people when everything else has failed. It works by getting participants to state a goal, having systematic group weigh-ins, and providing lots and lots of peer encouragement. The Paris Agreement is a far more serious proposition than that—the health of us all and the planet itself is at stake, along with trillions of dollars of investments. But despite its voluntary nature, one very important thing that Paris has done is to send a strident signal to industry that business is not going to be done as usual, because the steps to achieving each nation's pledge are public knowledge. In the case of the United States, for example, our 2015 nationally determined commitment (NDC) included a plan to tighten our standards for the average miles per gallon (or fuel economy) that cars and light trucks would have to meet in the future, providing a target for the American auto industry (and most were already willing to do so, given the need to compete with the rest of the world for the increasingly global auto market). It shouldn't be traumatic for America to meet these goals, despite its adoration of the automobile. Electric vehicles were once an expensive pipedream, but the recent collapse in the cost of batteries has made electrification of the auto industry so practical that even General Motors has announced a plan to produce only electric cars by 2035.[29] Even car-lovers are mollified when they consider that the transition will take place over years, just as the transition to catalytic converters to fight smog did. There will be a transition period as more and more new cars are sold with electric technology and charging stations are created to meet demand, but the old gasoline-powered cars will still be around for a while.

However, the Paris Agreement's promises of financial and technological transfers from wealthier nations to poorer ones in order to aide their transitions to clean energy have been slow to materialize. How much of the promised money will take the form of public investment (i.e., foreign aid) versus private investment by companies (with all the attendant dangers of exploitation in a for-profit regime) remains unspecified. A scenario in which large companies use the Paris Agreement to wangle lucrative contracts in sub-Saharan Africa that could reduce carbon emissions but would also increase poverty through price gouging would be a formula for intolerable global strife. Watchful attention as the agreement gains momentum will be focused on this danger and can probably avoid it.

The Paris Agreement didn't happen by chance. The timing was right because a practical version of a future without fossil fuels was starting to take shape, and the fantastic news is that it has rapidly snowballed since. The prices of solar and wind energy have been dropping rapidly over the past decade as technologies have improved and vast increases in use have brought costs under control. Almost everywhere in the world, it's now cheaper to build and operate a renewable-energy power plant than one using coal, oil, or even gas. These are the stunning new messages of hundreds of experts and the world's governments, who agreed on them for the first time in the 2022 report from the Intergovernmental Panel on Climate Change (IPCC). Which technology is the most economical for new power plants varies by region, depending on such factors as indigenous fossil fuel resources, labor prices, discount rates applied to costs over the plant's lifetime, and required safety and security measures (particularly for nuclear energy). Even in the fossil energy–rich United States, solar and onshore wind-based power plants were cheaper to construct and operate than those using methane gas by around 2016. Gas power plants in turn had been considerably cheaper than those using coal for at least seven years before that, and former president Trump's efforts to prop up America's failing coal industry floundered because coal plants were simply uneconomic. If you build a new power plant in the United States, you'll try to get the most watts for your dollars over the lifespan of the plant, and so the practical approach is to build a solar or onshore wind plant, or perhaps gas if your local solar and wind sources are intermittent. By 2020, US coal consumption had fallen by 40% since 2016 (although exports kept the loss of production to a smaller decrease, about 16%). US consumption of renewable energy surpassed coal use in 2019, before Trump even left office.

Predictably, fossil fuel supporters argue that sustainable mining and disposal of solar, wind, and battery materials is an insoluble problem. I have to laugh when I hear that, since it is the unbridled release of long-lived carbon dioxide waste from fossil fuels that is unsustainable. My view is that alternative energy materials will all need to be developed with sustainability and management requirements firmly in mind, and such ways to do better include very tough recycling rules. Alternative battery materials like sodium and iron are also under development.[30] I think we can avoid making the mistake of ignoring persistent waste again, now that we've seen it come back to haunt us so many times.

Holding climate change to less than 2°C above preindustrial values will require global greenhouse gas emissions to reach a net of zero by about 2050. This refers to what humanity puts in

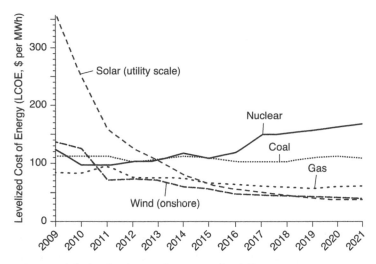

Estimates of the levelized cost of US energy for different sources since 2009. The comparison is for various generation technologies over the lifetime of the energy source on a dollars-per-MWh basis, including sensitivities for US federal tax subsidies, fuel prices, carbon pricing, and cost of capital.

minus added sinks—implying that any remaining sources are equaled by sinks (such as active removal mechanisms, which might include direct air capture of carbon, reforestation, and the like). Accounting accurately for these terms is a huge challenge and will require ongoing scrutiny. In 2020, the European Union was the first to promise to achieve net zero by 2050, followed closely by China's promise to follow suit by 2060, and the Biden campaign's vow to do the same as the European Union. People ask me if holding global warming to 2°C is possible. I can now tell them that not only is it possible, but, if these three major powers achieve their promises, then it will very likely happen.[31] Even a couple of years ago, I wouldn't have been able to say that, and people's concerns often come from out-of-date information on this fast-moving topic.

A warming of 2°C as a global average doesn't mean that our climate will not be altered, and it doesn't mean that India and other nations won't also need to reduce emissions over time as technologies become much cheaper. Warming will be considerably greater on land (by at least 50%), and we can expect the increase in the entire Arctic to be about twice as great as the global average. Sea levels will continue to rise, and wildfires will rage until many forest ecosystems are converted to scrublands or even deserts. Earth will be a different planet, and there will be human suffering. But if we hold global warming to 2°C or less, it will be a huge success compared to the 4°C pathway we were on before Paris. Since Paris, we have already dropped down from an expected 4°C future to about 3°C by 2100 if countries stay on track with their commitments—a remarkable change.

The 2050 promises made by governments are so far in the future that they are largely aspirational. What matters much more are near-term promises and actions, since it is these that are fast turning the crank of technology-forcing. Holding warming to 1.5°C remains a goal to strive for, although experts are becoming pessimistic that it can be achieved, and concern is growing about the 2°C target as well, which is all the more reason why the near term is essential. In July of 2021, both the European Union and the United States began announcing plans to put meat on the bones of their promises. The EU plans are bold. They include an end to diesel and gasoline car sales by 2035 and a carbon border tax on products such as steel and fertilizer, which will ensure that EU companies can compete and will put tremendous pressures on other nations to reduce their emissions.[32] President Biden is clearly determined to avoid restricting climate action to executive orders that could easily be overturned by his successor (as Obama's were), turning instead to the more long-lasting route of legislation. The Biden era Inflation Reduction act is already law, and contains an estimated $370 billion dollars of climate-related funding. While laws can be overturned by new congressional actions or Supreme Court decisions, that's a lot harder than a signature on an executive order. The act is expected to reduce US emissions by about 30%–40% by 2030, by such measures as financial support to companies that switch from fossil fuel based infrastructure to carbon-free sources, and tax credits to consumers and companies alike to foster use and development of clean energy equipment (including electric vehicles and heat pumps for example).[33] Regardless of whether these specific EU and US actions are hung up in court, fail, or languish in negotiations for a while, the technology steering signals to industry that they represent are clear: there will be change.

To be sure, some key issues remain to be fully addressed if the world is to fulfill the ambitions of Paris. As the share of renewables grows, the question of how to deal with potentially intermittent sunshine and wind increasingly rears its head. Batteries are still expensive. This means that a mix of municipal energy sources is needed for now, both in the United States and elsewhere (such as China), to handle the intermittency of Sun and wind, although battery costs are coming down. To maintain a dependable supply of electricity as solar and wind dominate the energy system will be a challenge that will require major changes at state and local levels that are only beginning to take shape. Processes such as issuing permits need a total revamp and should be a priority along with advances in battery storage technology. My friends who are experts on China tell me that the division between national goals and capabilities versus state and local implementation is also a key reason why coal use continues to increase in China. Improvements in grid systems (so-called smart grids) that can efficiently transfer more load over longer distances are also likely to be a part of a national solution. Some argue that distributed home solar, with batteries inside garages and such, will increasingly displace centralized municipal power plants in parts of the United States. Most experts look forward to an increasingly green and increasingly electrified world, with almost everything from home and industrial heating to power for cars, trucks, and trains switching over to electric technologies, so we also need to generate far more electricity, which is a huge task. Eventually even air travel will have to switch to non-fossil fuels, but most experts suggest that it will be one of the last businesses to do so. While its per capita footprint is large and expected to grow in the long term, not that many people fly very often. The total carbon emissions from aviation were only about a few percent of global emissions in 2020, implying that putting the near-term focus on other, more readily replaced sources makes practical sense.

Some politicians continue to howl that these proposals are far too expensive. But I think my MIT colleague, the respected environmental economist John Reilly, has it right when he says, "On the cost, I think we have to take it one step at a time." He argues that we should "focus on making the cuts we can now for a reasonable foreseeable future. . . . Ten years further down the road, it will be clearer what is possible in the next ten years." Paralysis now because we are not sure about the complete step-by-step route to the finish line is not only impractical, it's the worst thing we can do. What's needed instead is to recognize what it takes to produce successful environmental action, and fight for it to happen now.

Old and young leaders have emerged to help humanity through that fight. One morning in 2015, I found myself riveted to a document the likes of which I had never imagined reading, much less being inspired by: an encyclical by Pope Francis. He wrote, "Climate change is a global problem with grave implications. . . . Its worst impact will probably be felt by developing countries: . . . changes in climate, to which animals and plants cannot adapt, . . . [and] the livelihood of the poor, who are then forced to leave their homes, with great uncertainty for their future and that of their children. . . . Sadly, there is widespread indifference to such suffering. . . . Our lack of response to these tragedies involving our brothers and sisters points to the loss of that sense of responsibility for our fellow men and women upon which all civil society is founded."[34] While he surely had help writing this encyclical, it was clear to me that his was a heartfelt plea to display compassion for what climate change can do not just to Catholics but all people.

A colleague of mine says nothing will happen about climate change until young people get mad. This makes me think of the upside of the angry days of Vietnam, when the social movement it inspired played a key role in action to deal with America's smog crisis of the 1970s. It brought us the Clean Air Act and saved cities like Los Angeles from oblivion. In August of 2018, a fifteen-year-old Swedish schoolgirl named Greta Thunberg started spending her days sitting on the steps of the parliament building in Stockholm holding up a sign reading "School Strike for Climate." She inspired a movement that stunned the world. Within days, others joined her on those steps. Growing numbers of school strikes popped up elsewhere, and one of the benefits of the Internet was the ability to coordinate million-student strong marches in numerous cities around the world within a year.[35] Divestment movements have spread across college campuses, as students demand that universities remove fossil fuels from their endowment investments. The current young generation embraces climate change as an urgent cause, drawing in not only their parents but also increasing numbers of other adults. As in the 1970s, public engagement is growing exponentially as the young insist that we deal with this issue.

While Thunberg exhorts the young, veteran climate activist Bill McKibben exhorts those older than sixty to use their knowledge and resources to combat the climate problem. As I heard McKibben wisely emphasize (and as the women who helped save Los Angeles from smog showed in an earlier chapter of this book), one person's letter to the editor won't usually go very far. But people banding together in groups can be much more influential—for example, to protest or withdraw their resources from banks that are heavily invested in fossil fuels. Such groups as the Citizens Climate Lobby also offer a way for people of all ages to band together in making an impact across national, state, and local levels.

People and activist organizations are using the powerful tool of the courts to influence governments and industry, just as the Environmental Defense Fund did with persistent pesticides. A court in the Netherlands recently ruled that the Shell corporation must cut its emissions faster.[36] Several US states have sued Exxon, Chevron, Shell, BP, and others for lying to shareholders and consumers about climate change.[37] Sixteen Montana teenagers have sued their state to fulfill the promise in its constitution to ensure a "clean and healthful environment."[38] Even if these particular suits are dismissed or overturned, the attention and pressure they create are important. They surely contributed to the recent election of several new members of Exxon's board of directors who are viewed as independent and inclined to adjust the business strategy to include efforts to combat emissions,[39] and contribute as well to the increasing societal questioning of whether fossil fuel companies are a good investment. Other environmental groups that have effectively worked to pressure industries and various nations at different times include the Sunrise Movement, Greenpeace, Extinction Rebellion, and more. The Sunrise Movement is often credited with the pressure that propelled the remarkable climate measures in the United States' Infrastructure and Inflation Reduction Acts in 2021 and 2022.[40]

The climate issue has become part of a broader social justice movement sweeping across America that is long overdue. The tragic video of the death of George Floyd at the hands of a white police officer made the oppression of black Americans suddenly more perceptible for all, inspiring people to take to the streets in protest. As we have seen, everything from air pollution to lead from paint and pipes has impacted America's poor more than the rich, making environmental justice part and parcel of the current social justice movement. Research demonstrates that climate change damages will also fall disproportionately on America's poorer groups.[41] Both the recognition of a changed climate and the entrainment of environment as part of a much larger social movement are reminiscent of the 1960s, when several of the environmental successes of the past century blossomed.

The ingredients are all there for successfully solving the climate change challenge, just as humans have done for the other success stories described in this book. This doesn't mean we returned nature to anything resembling a pristine state, but we did manage our environmental impacts in a way that greatly reduced damages and risks.

Public engagement and climate change social movements have forced some politicians to act and allowed others to do what they already had in mind, leading to the creation of the innovative Paris Agreement, which accounts for the singular attributes of the climate problem and its deep connections to all world economies. The Kigali Amendment to the Montreal Protocol to reduce the hydrofluorocarbon contribution to climate change shows that it is certainly possible to enact effective national and even global policy to mitigate climate change. The only thing special about carbon dioxide is its scale, connections to economy, and the associated fear of its practical side: costs, particularly up-front costs. But in this as in other technological change, the up-front costs have come down at an astonishing rate and are expected to continue to do so. Technology-steering through many national and international policies has transformed a slow transition that was steadily advancing over many years into a rapid explosion in newly affordable alternative technologies to reduce our greenhouse gas emissions during the past decade. Policy that forces or steers technology change is essential in this problem, just as it was in Muskie's Clean Air Act and the game-changing catalytic converter that reduced smog ozone and saved millions if not billions of children not only from asthma but also from the poison of too much lead in their blood.

The clean energy revolution is now practical, and it is ultimately unstoppable. The playing field has changed now that renewable sources have become cheaper than fossil fuels and can power nearly all transportation. Transforming our systems from top to bottom is still a big task but the global transition is inevitable. Time is crucial if we are to safeguard the planet and ourselves.

It would have been better if politicians, industry, and people had exercised more caution related to climate change sooner, as carbon dioxide concentrations continued to increase and the world slowly warmed, but we've now reached the point where the climate has changed enough to become personal and perceptible to many people. That this occurred without the visible shroud of smog, the obvious deaths of birds and other wildlife from persistent pesticides, or the hot crisis of an ozone hole is a miracle of public awakening. In this respect climate change is perhaps most like the case of lead: we knew about the problem for centuries and yet managed to ignore it while the problem slowly got worse, until evidence for the poisoning became obvious and perceptible to everyone. That is all the more reason for hope in the future—because the planet will surely scream louder and louder for us to continue to pay attention as our climate keeps changing. The climate change challenge is indeed the biggest one that humanity has ever faced, and, like those of the past, it will only be conquered with public engagement. People in some segments of society will be hostile or uninterested, but most Americans today are concerned at some level about climate change, a statistic that has grown systematically in the past decade as climate changes became more personal and perceptible.[42] They are also generally supportive of alternative energy.[43]

There is no magic formula for the public activism that inspires political will and effective policy, and what will work depends very much on individuals and events. Greta Thunberg sat alone on the steps of the Swedish Parliament building with signs and inspired an influential movement that grew to involve over seven million people. Not everyone can be a leader or a hero. The playbook for what's needed is clear, and players of every kind are needed to make climate change solvable.

In January of 2007, I co-chaired the official plenary that approved the 2007 IPCC Scientific Assessment of Climate Change together with my colleague Qin Dahe of China. The approval session took place at a UN facility in Paris, and involved 113 countries set out in rows behind their national name plates, with Dahe and I up on the center of the dais flanked by several other scientists. I remember the hall as very crowded, with very high ceilings, dark benches, and poor lighting. It was the culmination of a six-year process of developing our massive report of over nine hundred pages with an author team of 152 international scientists from around the world, about 30 of whom were in Paris with us. Now we had to finalize the words of the Summary for Policymakers, and the so-called headline statements.

We had drafted the text that we'd present to the policymakers earlier and provided our rationale for the statements in careful presentations by the scientists. Painstakingly, we then proposed, and they approved each and every word after extensive comment and discussion. Dahe and I would let minor changes in as long as they didn't change the meaning of the science, and indeed I have always marveled at the ability of the diplomats to make the wording clearer. That is, when they want to—some will occasionally try to make it confusing if they don't like a particular point, but it is a joy to chair the process by which so many other nations never let them get away with that, because they all value strong science as the bedrock on which they can work, and they know it when they see it.

We had arrived at our biggest headline statement, the one we all knew would attract newspaper headlines. We scientists had wrestled with its wording. How do we convey the fact that we now have so much data from independent methods that all show the world is indeed warming, not only using thermometers on land at dozens of stations worldwide, but also from retreat of the world's glaciers, the warming of the world's oceans, documented retreat of snow cover, and rising sea levels? We scientists had tried out a dozen different constructions of this keystone sentence among ourselves in drafting sessions, all much too technical. Then we came up with the right scientific words: warming is unequivocal.

When we presented those three remarkable words on the screen for all to contemplate, a hush fell over the room. Switzerland raised its flag to comment. The diplomat used flowery terms to describe how important language was in his country, with its several national languages. He declared that unequivocal was OK in English but he was concerned that it wouldn't translate into French. France immediately raised its flag to curtly say that nothing would be lost in translation to French. Colombia immediately followed by voicing the same concern about Spanish, only to be quickly rebutted by Spain. The room fell silent again. Then I said, "Dear diplomats, may I suggest that we leave translation to the professionals who will perform that task for the whole document." When no more flags were raised and when I felt that the requisite pause was sufficient, I urgently whispered to Dahe to gavel it now, and he whacked the hammer down so hard I thought the handle would break. The room burst into spontaneous applause, and I realized that the train had left the station: the world would indeed make progress on climate change. The Rubicon had been crossed because the information from science and their own observations made climate change perceptible enough even for demanding people like these diplomats.

As we did for a full week, we worked long past the dinner hour, had a short break for some of the most tasteless sandwiches I've ever had in France, and continued our session late into the night. The Eiffel tower, that beautiful monument to technology as well as science, was illuminated with colored lights as I walked back to my hotel at 2:00 a.m. I stared at the tower a long time, marveling at it along with the events of the week. Greenpeace activists had climbed up on its graceful frame to place a huge banner commemorating our report that read "It's not too late." And now it's definitely time to hurry. But I'm more convinced than ever that they were right.

Greenpeace activists climbing more than three hundred feet (top) to place a banner (bottom) on the Eiffel Tower during the IPCC (2007) approval process.

ACKNOWLEDGMENTS

Solvable began as a solitary quest. But just as I found that healing the Earth invariably requires contributions from many different people, so too has the book itself only come to fruition because of many people who generously gave both their time and intellectual energy to help.

First, I am grateful to colleagues Judy Layzer, Harriet Ritvo, Janelle Knox-Hayes, and Kate Brown, who guided this physical scientist to a far deeper appreciation of history and the social sciences. I also deeply appreciate the friends, students, and colleagues who read drafts of the chapters and immeasurably contributed to their improvement, including Barbara Beatty, Kate Brown, Alice Cort, Dale Durran, Cass Grenier, Megan Lickley, Harriet Ritvo, Alan Robock, and Kasturi Shah. The many students who took my classes over the years also deserve thanks for their questions, shared viewpoints, and thoughtful criticisms, which improved my thinking and this book. I also acknowledge with deepest thanks my writers' group, who patiently listened to, commented upon, and vastly improved more than one draft: Stephanie Bendel, John Christenson, Judy Gilligan, David Kline, Barbara Shark, and Joe White. I also thank my agent Carolyn Savarese and her husband and sidekick, William Patrick, for their many helpful suggestions. I am deeply grateful to my editorial assistant, Geraldine McGowan, for a broad range of assistance, without which this book would not have been possible. My editor Joe Calamia, Matt Lang, and the rest of the University of Chicago team expertly sharpened the text, illustration choices, and other components, and I am extremely grateful to them. Finally, I thank my husband Barry Sidwell and stepson Casey Sidwell for many hours of helpful conversation about the book and patient forbearance with my obsession.

NOTES

CHAPTER ONE

1. G. P. Brasseur, *The Ozone Layer: From Discovery to Recovery* (Boston: American Meteorological Society, 2019), ix–x, 1–3.
2. Brasseur, *Ozone Layer*, 47–59.
3. L. Greenhouse, "Protest Blossoms as Sonic Booms," *Harvard Crimson*, September 26, 1967; https://www.thecrimson.com/article/1967/9/26 /protest-blossoms-as-sonic-booms-psix/.
4. P. M. Morrisette, "The Evolution of Policy Responses to Stratospheric Ozone Depletion," *Natural Resources Journal* 29 (1989): 793–820.
5. Morrisette, "Evolution of Policy Responses."
6. H. S. Johnston, "Atmospheric Chemistry Research at Berkeley," Oral history conducted in 1999 by Sally Smith Hughes, PhD Regional Oral History Office, The Bancroft Library, University of California, Berkeley (2005); https://digitalassets.lib.berkeley.edu/roho/ucb/text/johnston _harold.pdf.
7. Johnston, "Atmospheric Chemistry Research."
8. Morrisette, "Evolution of Policy Responses."
9. J. E. Lovelock and R. J. Maggs, "Halogenated Hydrocarbons in and over the Atlantic," *Nature* 241 (1973): 194–96.
10. Brasseur, *Ozone Layer*, 172–73.
11. M. J. Molina and F. S. Rowland, "Stratospheric Sink for Chlorofluoro-methanes: Chlorine Atom-Catalyscd Destruction of Ozone," *Nature* 249 (1974): 810–12.
12. L. Dotto and H. Schiff, *The Ozone War* (New York: Doubleday, 1978), 146–49.
13. Dotto and Schiff, *Ozone War*.

14. S. Roan, *Ozone Crisis: The 15-Year Evolution of a Sudden Global Emergency* (New York: Wiley, 1990), 57.

15. Johnston, "Atmospheric Chemistry Research."

16. Roan, *Ozone Crisis.*

17. R. Benedick, *Ozone Diplomacy* (Cambridge, MA: Harvard University Press, 1991), 27.

18. S. C. Johnson and Company, "Taking CFCs Out of Aerosols: How Sam Johnson Led S. C. Johnson to Environmental Activism"; https://www .scjohnson.com/en/about-us/the-johnson-family/sam-johnson/taking -cfcs-out-of-aerosols-how-sam-johnson-led-sc-johnson-to-environ mental-activism.

19. "An Aerosol Ban in Force Today in Oregon," *New York Times*, March 1, 1977; https://www.nytimes.com/1977/03/01/archives/an-aerosol-ban-in -force-today-in-oregon-an-aerosol-ban-in-force-in.html.

20. E. M. Leeper, "Spray Cans Must Go," *Bioscience* 25 (1975): 753–55.

21. "Most Aerosols Face a Swedish Ban," *New York Times*, January 30, 1978; and Benedick, *Ozone Diplomacy*, 23–27.

22. J. M. Turner and A. S. Isenberg, *The Republican Reversal: Conservatives and the Environment from Nixon to Trump* (Cambridge, MA: Harvard University Press, 1988).

23. Roan, *Ozone Crisis.*

24. Personal communication from Dr. Joseph Steed, May 13, 2021.

25. W. Prochnau, "The Watt Controversy," *Washington Post*, June 30, 1981; https://www.washingtonpost.com/archive/politics/1981/06/30/the-watt -controversy/d591699b-3bc2–46d2–9059-fb5d2513c3da/; and "Car Emis- sion Rule Relief," *New York Times*, October 6, 1981.

26. F. X. Clines, "Watt Asks That Reagan Forgive 'Offensive' Remark about Panel," *New York Times*, September 23, 1983; https://www.nytimes .com/1983/09/23/us/watt-asks-that-reagan-forgive-offensive-remark -about-panel.html.

27. National Research Council, *Causes and Effects of Changes in Stratospheric Ozone: Update 1983*, Washington, DC: National Academies Press, 1984; https://doi.org/10.17226/19330.

28. Morrisette, "Evolution of Policy Responses," and Benedick, *Ozone Diplomacy*, 40–47.

29. See S. Solomon, "Stratospheric Ozone Depletion: A Review of Con- cepts and History," *Reviews of Geophysics* 37 (1999): 275–316 and refer- ences therein.

30. Solomon, "Stratospheric Ozone Depletion."

31. C. Peterson, "Chlorofluorocarbon Group Supports Production Curbs," *Washington Post*, September 17, 1986; https://www.washingtonpost.com /archive/politics/1986/09/17/chlorofluorocarbon-group-supports-produc tion-curbs/dd975cc0-b02b-4e08–9aef-b0711e6eab70/.

32. Benedick, *Ozone Diplomacy*, 55–117.
33. Benedick, *Ozone Diplomacy*, 59–67.
34. Benedick, *Ozone Diplomacy*, 91–10;1
35. https://www.congress.gov/treaty-document/100th-congress/10/reso lution-text.
36. See the *New York Times* website: https://www.nytimes.com/1988/03/26 /business/behind-du-pont-s-shift-on-loss-of-ozone-layer.html.
37. Benedick, *Ozone Diplomacy*, 114.
38. See Solomon, "Stratospheric Ozone Depletion."
39. Benedick, *Ozone Diplomacy*, 118–96.
40. S. Solomon, D. J. Ivy, D. Kinnison, M. Mills, R. R. Neely III, and A. Schmidt, "Emergence of Healing in the Antarctic Ozone Layer," *Science* 353 (2016): 279–74.
41. *Scientific Assessment of Ozone Depletion: 2022*, Geneva: World Meteorological Organization, GAW report 278.

CHAPTER TWO

1. V. Ramanathan, "Greenhouse Effect Due to Chlorofluorocarbons: Climatic Implications," *Science* 190 no. 4209 (1975): 50–52.
2. S. Barrett, "Political Economy of the Montreal Protocol," *Oxford Rev. Econ. Pol* 14 (1999): 20–39.
3. J. Rees, "I Did Not Know . . . Any Danger Was Attached: Safety Consciousness in the Early American Ice and Refrigeration Industries," *Technology and Culture* 46, no. 3 (2005): 541–60.
4. G. C. Briley, "A History of Refrigeration," *ASHRAE Journal* 46, no. 11 (2004): S31–S34.
5. EUR-LEX, access to European Union Law, "Emissions from Air-Conditioning Systems in Motor Vehicles"; https://eur-lex.europa.eu /EN/legal-content/summary/emissions-from-air-conditioning-systems -in-motor-vehicles.html.
6. M. S. Reisch, "Earth-Friendly Cool Cars in 2011: Fluorine-Based-Refrigerant Makers Jockey to Save Auto-Air-Conditioning Market from CO_2 Threat," *Chemical & Engineering News* 84, no. 36 (2006): 26–28.
7. Intergovernmental Panel on Climate Change (IPCC), *Safeguarding the Ozone Layer and the Global Climate* (Cambridge: Cambridge University Press, 2005), special report, ISBN 92-9169-118-6.
8. G. J. M. Velders, S. O. Andersen, J. S. Daniel, D. W. Fahey, and M. McFarland, "The Importance of the Montreal Protocol in Protecting Climate," *Proceedings of the National Academy of Sciences* 104 (2007): 4814–19.

9. G. J. M. Velders, D. W. Fahey, J. S. Daniel, M. McFarland, and S. O. Andersen, "The Large Contribution of Projected HFC Emissions to Future Climate Forcing," *Proceedings of the National Academy of Sciences* 106 (2009): 10949–54.

10. FAO (Food and Agriculture Organization of the UN), *Global Food Losses and Food Waste: Extent, Causes and Prevention* (Rome: FAO, 2011); see also FAO, *Save Food for a Better Climate* (Rome: FAO, 2017).

11. Champions 12.3, "Sustainable Development Goals (SDG) Target 12.3 on Food Loss and Waste: 2021 Progress Report," World Resources Institute and Ministry of Agriculture Nature and Food Quality of the Netherlands, 2021. https://champions123.org/publication/sdg-target -123-food-loss-and-waste-2021-progress-report; and Rovshen Ishangu-lyyev, Sanghyo Kim, and Sang Hyeon Lee, "Understanding Food Loss and Waste: Why Are We Losing and Wasting Food?" *Foods* 8, no. 8 (2019): 297; https://doi.org/10.3390/foods8080297.

12. K. Fay, personal communication to the author, 2021.

13. M. W. Roberts, "Finishing the Job: The Montreal Protocol Moves to Phase Down Hydrofluorocarbons," *Review of European Comparative & International Environmental Law* 26, no. 30 (2017): 220–30.

14. "Xi, Obama Meet for First Summit," *China.Org.Cn.*, June 8, 2013; http:// www.china.org.cn/world/2013–06/08/content_29067028.htm.

15. See https://obamawhitehouse.archives.gov/the-press-office/2013/06/08 /united-states-and-china-agree-work-together-phase-down-hfcs.

16. See https://obamawhitehouse.archives.gov/the-press-office/2014/09/30 /us-India-joint-statement.

17. Roberts, "Finishing the Job."

18. See https://www.ecacool.com/en/publications/kerry_remarks_mon treal_protocol/.

19. "U.S. Enacts HFC Phase-Down Law as Part of COVID Relief Bill"; https://coolcoalition. org/u-s-enacts-hfc-phase-down-law-as-part-of-covid-relief-bill/.

20. S. Mufson, "Climate Solutions: U.S. Ratifies Global Treaty Curbing Cli- mate Super-Pollutants," *Washington Post*, September 21, 2022; https:// www.washingtonpost.com/climate-solutions/2022/09/21/kigali-amend ment-senate-super-pollutants-climate/.

CHAPTER THREE

1. Steve Swatt, Susie Swatt, R. Lavally, and J. Raimundo, "Activist Women Kickstart California's Conservation and Clean Air Movements," in *Paving the Way: Women's Struggle for Political Equality in California* (Berkeley, CA: Berkeley Public Policy Press, 2019).

2. Steve Swatt and Susie Swatt, "The Women behind California's Landmark Vehicle Emissions Law"; https://calmatters.org/commen tary/2019/10/california-emissions-law/ (2019).

3. As seen in the photo by H. O'Neil, "Liberty Enlightening the World— Inauguration of the Bartholdi Statue," Library of Congress, LC-DIG-ds-04491; http://loc.gov/pictures/resource/cph.3a55010/.

4. L. Makra and P. Brimblecombe, "Selections from the History of Environmental Pollution, with Special Attention to Air Pollution," part 1, *International Journal of Environment and Pollution* 22 (2004): 641–56.

5. K. C. Heidorn, "A Chronology of Important Events in the History of Air Pollution Meteorology to 1970," *Bulletin of the American Meteorological Society* 59 (1978), 1589–97.

6. Heidorn, "History of Air Pollution Meteorology."

7. R. Stone, "Counting the Cost of London's Killer Smog," *Science* 298 (2002): 2106–7.

8. Heidorn, "History of Air Pollution Meteorology."

9. A. C. Stern, "History of Air Pollution Legislation in the United States," *Journal of the Air Pollution Control Association* 32 (1982): 44–61.

10. L. P. Snyder, "The Death-Dealing Smog over Donora, Pennsylvania: Industrial air Pollution, Public Health Policy, and the Politics of Expertise, 1948–1949," *Environmental History Review* 118 (1994): 117–39.

11. Snyder, "Death-Dealing Smog."

12. Stern, "History of Air Pollution Legislation."

13. P. J. Lioy and P. G. Georgopoulos, "New Jersey: A Case Study of the Reduction in Urban and Suburban Air Pollution from the 1950s to 2010," *Environmental Health Perspectives* 19 (2011): 1351–55.

14. D. Vogel, "Protecting Air Quality," in *California Greenin': How the Golden State Became an Environmental Leader* (Princeton, NJ: Princeton University Press (2018).

15. Vogel, "Air Quality."

16. Vogel, "Air Quality."

17. Vogel, "Air Quality."

18. Vogel, "Air Quality."

19. J. N. Pitts Jr. and E. R. Stephens, "The Pioneers," *Journal of the Air Pollution Control Association* 28 (1978): 516–17.

20. Arnold O. Beckman, Interview by Mary Terrall, October 16–December 4, 1978, Oral History Project, California Institute of Technology Archives.

21. Beckman, quoted in Zus Haagen-Smit, Interview by Shirley K. Cohen, Pasadena, California, March 16 and 20, 2000; Oral History Project, California Institute of Technology Archives.

22. Beckman, quoted in Zus Haagen-Smit, Interview by Shirley K. Cohen.

23. A. J. Haagen-Smit, E. F. Darley, M. Zaitlin, H. Hull, and W. Noble, "Investigation on Injury to Plants from Air Pollution in the Los Angeles Area," *Plant Physiology* 27 (1952): 18–34.
24. Haagen-Smit et al., "Injury to Plants from Air Pollution."
25. C. E. Bradley and A. J. Haagen-Smit, "The Application of Rubber in the Quantitative Determination of Ozone," *Rubber Chemistry and Technology* 24 (1951): 750–55.
26. Bradley and Haagen-Smit, "Application of Rubber."
27. H. S. Johnston, "Atmospheric Chemistry Research at Berkeley," Oral history conducted in 1999 by Sally Smith Hughes, PhD, Regional Oral History Office, The Bancroft Library, University of California, Berkeley.
28. A. J. Haagen-Smit., C. E. Bradley, and M. M. Fox, "Ozone Formation in Photochemical Oxidation of Organic Substances," *Industrial and Engineering Chemistry* 45 (1953): 2086–89.
29. Johnston, "Atmospheric Chemistry Research."
30. Johnston, "Atmospheric Chemistry Research."
31. J. Bonner, "Arie J. Haagen-Smit: A Biographical Memoir," *Biographical Memoirs of the National Academy of Sciences* 53 (1989): 187–216.
32. Swatt, Swatt, Lavally, and Raimundo, "Activist Women Kickstart California's Conservation and Clean Air Movements," in *Paving the Way*.
33. Vogel, "Air Quality."
34. R. Popkin, "Environmental Annals: Two 'Killer' Smogs the Headlines Missed," *EPA Journal* 12 (1986): 27–38.
35. E. A. Asbury, "Smog Is Really Smaze; Rain May Rout It Tonight," *New York Times*, November 21, 1953.
36. J. K. Goldstein and Edmund S. Muskie, "The Environmental Leader and Champion," *Maine Law Review* 67 (2015): 226–32.
37. E. Langer, "Pollution Politics: LBJ Retreats on Opposition to Measure Curbing Pollution from Automobile Exhaust," *Science* 148 (1965): 611–13.
38. Lyndon B. Johnson, "The Great Society Speech," transcript of speech delivered at the University of Michigan, May 22, 1964. MichiganOnline by Gerhard Peters and John T. Woolley, *The American Presidency Project*; https://www.presidency.ucsb.edu/node/239689.
39. A. Norwood, "Claudia 'Lady Bird' Johnson," National Women's History Museum, 2017; www.womenshistory.org/education-resources/biographies/claudia-lady-bird-johnson.
40. Lady Bird Johnson, "Quotes," The Lady Bird Wildflower Center, University of Texas, Austin; http://www.ladybirdjohnson.org/quotes.
41. A. Rome, "Give Earth a Chance: The Environmental Movement and the Sixties," *Journal of American History* 90 (2003): 525–54.

42. L. Thulin, "How an Oil Spill Inspired the First Earth Day," *Smithsonian Magazine*, April 22, 2019; https://www.smithsonianmag.com/history /how-oil-spill-50-years-ago-inspired-first-earth-day-180972007/.

43. M. Rinde, "Richard Nixon and the Rise of American Environmentalism," *Distillations*, June 2, 2017; https://www.sciencehistory.org/distillations /richard-nixon-and-the-rise-of-american-environmentalism.

44. J. Latson, "The Burning River That Sparked a Revolution," *Time*, June 22, 2015; https://time.com/3921976/cuyahoga-fire/#:~:text=The %20flaming%20Cuyahoga%20became%20a,and%20state%20 environmental%20protection%20agencies.

45. D. Zierler, *The Invention of Ecocide: Agent Orange, Vietnam, and the Scientists Who Changed the Way We Think about the Environment* (Athens: University of Georgia Press, 2011).

46. Rome, "Give Earth a Chance."

47. Rome, "Give Earth a Chance."

48. Richard Nixon, "Annual Message to the Congress on the State of the Union," January 20, 1970.

49. "Muskie Hails Nixon's 'Rhetoric of Concern' on Pollution But Wants Specifics," *New York Times*, January 24, 1970.

50. E. Muskie on *Meet the Press*, February 1, 1970. https://scarab.bates.edu /maf/32/

51. E. Muskie 1970 speech index, Papers of Edmund S. Muskie, Bates College; https://www.bates.edu/archives/files/2017/03/MuskieSpeechIndex 1970toJanuary1981.pdf.

52. E. M. Muskie, "Our Polluted America: What Women Can Do," *Ladies' Home Journal*, February 1970 (also available in US *Congressional Record* of February 25, 1970; Rep. Roman Pucinski read it into the record).

53. Rome, "Give Earth a Chance."

54. A. Rome, *The Genius of Earth Day: How a 1970 Teach-in Unexpectedly Made the First Green Generation* (New York: Hill and Wang, 2013).

55. H. Erskine, "The Polls: Pollution and Its Costs," *Public Opinion Quarterly* 36 (1972): 120–35.

56. O. B. Waxman, "Meet 'Mr. Earth Day,' the Man Who Helped Organize the Annual Observance," *Time*, April 19, 2019; https://time.com/5570 269/earth-day-origins/.

57. R. L. Ash, G. P. Baker, J. B. Connally, F. R. Kappel, R. M. Paget, and W. E. N. Thayer, "Ash Council (Executive Office of the President, President's Advisory Council on Executive Organization) Memorandum to the President, April 29, 1970," US Environmental Protection Agency Archive; https://www.epa.gov/archive/epa/aboutepa/ash-council-memo .html.

58. R. Cohen and M. Koncewicz, "Tin Soldiers and Nixon's Coming: The Shootings at Kent State and Jackson State at 50," *Nation*, May 4, 2020.
59. L. Billings, "Oral History." Interview by Don Nicoll, Bates College SCARAB. September 16, 2002.
60. P. G. Rogers, "The Clean Air Act of 1970," *EPA Journal* 16 (1990): 21–23.
61. Billings, "Oral History."
62. A. E. Carlson, "California Motor Vehicle Standards and Federalism: Lessons for the European Union," Univ.of California–Berkeley Working Papers (2008); https://escholarship.org/uc/item/9fr4j8rf.
63. Bonner, "Arie J. Haagen-Smit."
64. A. J. Haagen-Smit, "A Lesson from the Smog Capital of the World," *Proceedings of the National Academy of Sciences* 67 (1970): 887–97.
65. D. Gerard and L. B. Lave, "Implementing Technology Forcing Policies: The 1970 Clean Air Act Amendments and the Introduction of Advanced Automotive Emissions Controls in the United States," *Technological Forecasting and Social Change* 72 (2005): 761–78.
66. Billings, "Oral History."
67. B. Daniels, A. P. Follett, and J. Davis, "The Making of the Clean Air Act," *Hastings Law Journal* 71 (2020): 901–58.
68. E. W. Kenworthy, "Tough New Clean-Air Bill Passed by Senate, 73–0," *New York Times*, September 22, 1975.
69. Daniels et al., "The Making of the Clean Air Act."
70. Billings, "Oral History."
71. J. M. Naughton, "President Signs Bill to Cut Auto Fumes 90% by 1977," *New York Times*, December 31, 1971.
72. E. M. Muskie, quoted in L. G. Billings, "Edmund S. Muskie: A Man with a Vision," *Maine Law Review* 67 (2015): 233–38.
73. W. D. Ruckelshaus, "Oral History Interview," interview by Dr. Michael Gorn, January 1993, *United States Environmental Protection Agency* 202-K-92–0003, 1993.
74. Ruckelshaus, "Oral History Interview."
75. Congressional Research Service Report, RL30853, "Clean Air Act: A Summary of the Act and Its Major Requirements," updated September 13, 2022; https://crsreports.congress.gov/search/#/?termsToSearch=308 53&orderBy=Relevance.
76. Gerard and Lave, "Implementing Technology-Forcing Policies."
77. Gerard and Lave, "Implementing Technology-Forcing Policies."
78. Gerard and Lave, "Implementing Technology-Forcing Policies."
79. Gerard and Lave, "Implementing Technology-Forcing Policies."
80. Gerard and Lave, "Implementing Technology-Forcing Policies."
81. L. Lundqvist, "Who Is Winning the Race for Clean Air? An Evaluation of the Impacts of the US and Swedish Approaches to Air Pollution Control," *Ambio* 8 (1979): 144–51.

82. Gerard and Lave, "Implementing Technology-Forcing Policies."
83. J. Lee, F. M. Veloso, D. A. Hounshell, and E. S. Rubin, "Forcing Technological Change: A Case of Automobile Emissions Control Technology Development in the US," *Technovation* 30 (2010): 249–64.
84. Billings "Oral History."
85. D. Calef and R. Goble, "'The Allure of Technology: How France and California Promoted Electric and Hybrid Vehicles to Reduce Urban Air Pollution," *Policy Science* 40 (2007): 1–34.
86. Calef and Goble, "Allure of Technology."
87. D. Kann, "Los Angeles Has Notoriously Polluted Air but Right Now It Has Some of the Cleanest of Any Major City"; https://www.cnn.com/2020/04/07/us/los-angeles-pollution-clean-air-coronavirus-trnd/index.html.
88. Calef and Goble, "Allure of Technology."
89. "Forcing Technology: The Clean Air Act Experience," *Yale Law Journal* 88 (1979): 1713–34; https://doi.org/10.2307/795694.
90. F. C. Menz and H. M. Seip, "Acid Rain in Europe and the United States: An Update," *Environmental Science & Policy* 7, 253–265 (2004).
91. M. E. Kraft and N. J. Vig, "Environmental Policy in the Reagan Presidency," *Political Science Quarterly* 99 (1984): 415–39.
92. R. E. Dunlap, "An Enduring Concern," *Public Perspective*, October 2002, 10–14.
93. P. Firozi, "How George H. W. Bush Helped Turn Acid Rain into a Problem of Yesteryear," *Washington Post*, December 4, 2018.
94. R. Conniff, "The Political History of Cap and Trade," *Smithsonian*, August 2009.
95. Connif, "History of Cap and Trade."
96. R. Schmalensee and R. N. Stavins, "The SO_2 Allowance Trading System: The Ironic History of a Grand Policy Experiment," *Journal of Economic Perspectives* 27 (2013): 103–22.
97. J. L. Stoddard, D. S. Jeffries, A. Lükewille, et al., "Regional Trends in Aquatic Recovery from Acidification in North America and Europe," *Nature* 401 (1999): 575–78.
98. Schmalensee and Stavins, "SO_2 Allowance Trading System."
99. A. Denny Ellerman and Juan-Pablo Montero, "The Declining Trend in Sulfur Dioxide Emissions: Implications for Allowance Prices," *Journal of Environmental Economics and Management* 36 (1998): 26–45.
100. Schmalensee and Stavins, "SO_2 Allowance Trading System."
101. Lundqvist, "Who Is Winning the Race for Clean Air?"
102. M. H. Lane, R. Morello-Frosch, J. D. Marshall, and J. S. Apte, "Historical Redlining Is Associated with Present Day Air Pollution Disparities in US Cities," *Environmental Science & Technology Letters* 9 (2022): 345–50.

103. M. L. Miranda, S. E. Edwards, M. H. Keating, and C. J. Paul, "Making the Environmental Justice Grade: The Relative Burden of Air Pollution Exposure in the United States," *International Journal of Environmental Research and Public Health* 8, (2011): 1755–71.

104. "Mexico City Cleans Up Its Reputation for Smog," *NBC News*, December 28, 2008; https://www.nbcnews.com/id/wbna28391130.

CHAPTER FOUR

1. J. Fernandez, "Help Save Your Children: On the Young Lords' Battle against Lead Poisoning in New York," *Lapham's Quarterly*, February 19, 2020; https://www.laphamsquarterly.org/roundtable/help-save-your-children.

2. J. Fernandez, *The Young Lords: A Radical History* (Chapel Hill, NC: University of North Carolina Press, 2020).

3. C. Warren, *Brush with Death: A Social History of Lead Poisoning* (Baltimore, MD: Johns Hopkins University Press, 2001).

4. Fernandez, "Help Save Your Children."

5. B. Berney, "Round and Round It Goes: The Epidemiology of Childhood Lead Poisoning, 1950–1990," *Milbank Quarterly* 71 (1993): 3–39.

6. D. Rosner and G. A. Markowitz, "A 'Gift of God'? The Public Health Controversy over Leaded Gasoline During the 1920s," *American Journal of Public Health* 75 (1985): 344–52.

7. Rosner and Markowitz, "A 'Gift of God'?"

8. M. A. Lessler, "Lead and Lead Poisoning from Antiquity to Modern Times," *Ohio Journal of Science* 88 (1988): 78–84.

9. S. Hong, J-P. Candelone, C. C. Patterson, and C. F. Boutron, "Greenland Ice Evidence of Hemispheric Lead Pollution Two Millennia Ago By Greek and Roman Civilizations," *Science* 265 (1994): 1841–44.

10. H. A. Waldron, "Lead Poisoning in the Ancient World," *Medical History* 17 (1973): 391–99.

11. Lessler, "Lead and Lead Poisoning."

12. G. Baker, "An Essay Concerning the Cause of the Endemial Colic of Devonshire, Which Was Read in the Theatre of the College of Physicians, in London, on the Twenty-ninth Day of June, 1776"; http://www.medicine.mcgill.ca/epidemiology/hanley/minimed/EssayConcerning CauseOfEndemialColicOfDevonshire.pdf.

13. "FDA Proposes Lead-Soldered Cans Be Banned," *Washington Post*, June 22, 1993; https://www.washingtonpost.com/archive/politics/1993/06/22/fda-proposes-lead-soldered-cans-be-banned/7ff9d1fa-c049-4498-b377-48a522c3cf92/.

14. D. Rosner, G. Markowitz, and B. Lanphear, "J. Lockhart Gibson and the Discovery of Lead Pigments on Children's Health: A Review of a Century of Knowledge," *Public Health Reports* 120 (2005): 296–300.
15. "Poison Peril Widespread," *New York Times*, December 31, 1924.
16. J. L. Gibson, "A Plea for Painted Railings and Painted Walls of Rooms as the Source of Lead Poisoning amongst Queensland Children," 1904 paper reprinted in *Public Health Reports* 120 (2005): 301–4.
17. A. Hamilton, "Lead Poisoning in the United States," *American Journal of Public Health* 4 (1914): 477–80; reprinted in 2009 in the same journal (vol. 99: S547–S549) after a US stamp was issued in her honor.
18. A. Hamilton, "What Price Safety: Tetra-ethyl Lead Reveals a Flaw in Our Defenses," *Journal of Occupational Medicine* 14 (1972): 98–100.
19. See the Harvard website; https://faculty.harvard.edu/dr-alice-hamilton.
20. J. Heitmann, "The ILO and the Regulation of White Lead in Britain During the Inter-war Years: An Examination of International and National Campaigns in Occupational Health," *Labor History Review* 69 (2004): 267–84.
21. The agreement is available at https://www.ilo.org/dyn/normlex/en/f?p=NORMLEXPUB:12100:0::NO::P12100_ILO_CODE:C013. For a list of countries that have ratified it, see https://www.ilo.org/dyn/normlex/en/f?p=1000:11300:0::NO:11300:P11300_INSTRUMENT_ID:312158.
22. Heitmann, "Regulation of White Lead in Britain."
23. Heitmann, "Regulation of White Lead in Britain."
24. J. Ludden, "Baltimore Struggles to Protect Children from Lead Paint"; https://www.npr.org/sections/health-shots/2016/03/21/471267759/baltimore-struggles-to-protect-children-from-lead-paint.
25. R. Rabin, "Warnings Unheeded: A History of Child Lead Poisoning," *American Journal of Public Health* 79 (1989): 1668–74.
26. R. A. Kehoe, F. Thamann, and J. Cholak, "Normal Absorption and Excretion of Lead," *Journal of the American Medical Society* 104 (1935): 90–92.
27. Clair Patterson, interview by S. K. Cohen, Pasadena, California, March 5, 6, and 9, 1995. Oral History Project, California Institute of Technology Archives.
28. Patterson, interview by S. K. Cohen.
29. Patterson, interview by S. K. Cohen.
30. Patterson, interview by S. K. Cohen.
31. Patterson, interview by S. K. Cohen.
32. C. Patterson, "Contaminated and Natural Lead Environments of Man," *Archives of Environmental Health* 11 (1965): 344–60.
33. *Hearings before a Subcommittee on Air and Water Pollution of the Senate Committee on Public Works*, 89th Cong., 2nd sess. (on S. 3112 . . . S.

3400 ... and contamination of the environment from lead and other substances, June 7, 8, 9, 14, and 15, 1966).

34. *Hearings before a Subcommittee on Air and Water Pollution.*

35. E. Muskie, "Clean Air Act Amendments of 1966," Muskie *Congressional Record*: CAA Debate, July 12, 1966; available online at http://abacus .bates.edu/muskie-archives/ajcr/1966/CAA%20Debate.shtml.

36. Muskie, "Clean Air Act Amendments of 1966."

37. D. M. Settle and C. C. Patterson, "Lead in Albacore: Guide to Lead Pollution in Americans," *Science* 207 (1980): 1167–76.

38. Patterson, interview by S. K. Cohen.

39. "EPA Takes Final Step in Phaseout of Leaded Gasoline," EPA press release, January 29, 1996; https://www.epa.gov/archive/epa/aboutepa /epa-takes-final-step-phaseout-leaded-gasoline.html.

40. K. Bridbord and D. Hanson, "A Personal Perspective on the Initial Federal Health-Based Regulation to Remove Lead from Gasoline," *Environmental Health Perspectives* 117 (2009): 1195–1201.

41. D. C. Bellinger and A. M. Bellinger, "Childhood Lead Poisoning: The Torturous Path from Science to Policy," *Journal of Clinical Investigation* 116 (2006): 853–57.

42. Bridbord and Hansen, "Health Policy."

43. Berney, "Epidemiology of Childhood Lead Poisoning."

44. Berney, "Epidemiology of Childhood Lead Poisoning."

45. Rabin, "Warnings Unheeded."

46. P. Mushak and A. F. Crocetti, "Methods for Reducing Lead Exposure in Young Children and Other Risk Groups: An Integrated Summary of a Report to the U.S. Congress on Childhood Lead Poisoning," *Environmental Health Perspectives* 89 (1990): 125–35.

47. Berney, "Epidemiology of Childhood Lead Poisoning."

48. "Nine Companies Penalized for Selling Children's Products That Violated the Federal Lead Paint Ban," Press Release, Consumer Products Safety Commission July 9, 2009; https://www.cpsc.gov/Newsroom /News-Releases/2009/Nine-Companies-Penalized-for-Selling-Child rens-Products-that-Violated-the-Federal-Lead-Paint-Ban.

49. Bridbord and Hansen, "Health Policy."

50. Bellinger and Bellinger, "Childhood Lead Poisoning."

51. Rabin, "Warnings Unheeded."

52. R. K. Byers, and E. E. Lord, "Late Effects of Lead Poisoning on Mental Development," *American Journal of Diseases of Children* 66 (1943): 471–83.

53. E. K. Silbergeld, "Preventing Lead Poisoning in Children," *Annual Reviews of Public Health* 18 (1997): 187–210.

54. H. Needleman, "Lead Poisoning," *Annual Review of Medicine* 55 (2004): 209–22.

55. Bellinger and Bellinger, "Childhood Lead Poisoning."

56. "Blood Lead Levels—United States, 1988–1991," Centers for Disease Control MMWR Weekly 43, no. 30 (1994): 545–48; https://www.cdc.gov /mmwr/preview/mmwrhtml/00032080.htm.

57. V. M. Thomas, R. H. Socolow, J. J. Fanelli, and T. G. Spiro, "Effects of Reducing Lead in Gasoline: An Analysis of the International Experience," *Environmental Science and Technology* 33 (1999): 3942–48.

58. "Inside the 20-Year Campaign to Rid the World of Leaded Fuel"; https://www.unep.org/news-and-stories/story/inside-20-year-cam paign-rid-world-leaded-fuel.

59. D. E. Jacobs, et al., The Prevalence of Lead-Based Paint Hazards in U.S. Housing," *Environmental Health Perspectives* 110 (2002): 599–606.

60. D. C. Bellinger, "Lead Contamination in Flint: An Abject Failure to Protect Public Health," *New England Journal of Medicine* 374 (2016): 1101–3.

61. See https://www.whitehouse.gov/brief ing-room/statements-releases/2021/12/16/ fact-sheet-the-biden-harris-lead-pipe-and-paint-action-plan/.

62. "Voters Support Fully Funding Lead Pipe Removal"; https://www.data forprogress.org/blog/2021/9/15/voters-support-fully-funding-lead -pipe-removal.

CHAPTER FIVE

1. O. Huckins, Letter to Rachel Carson, Rachel Carson Papers (YCAL MSS 46), Beinecke Rare Book and Manuscript Library, Yale University; https://collections.library.yale.edu/catalog/2026498.

2. R. Carson, "Acknowledgments," in *Silent Spring* (Boston: Houghton Mifflin, 1962).

3. E. Angelakis, Y. Bechah, and D. Raoult, "The History of Epidemic Typhus," *Microbiology Spectrum* 4, no. 4 (2016).

4. D. Raoult, O. Dutour, L. Houhamdi, R. Jankauskas, P-E. Fournier, Y. Ardagna, M. Drancourt, et al., "Evidence for Louse-Transmitted Diseases in Soldiers of Napoleon's Grand Army in Vilnius," *Journal of Infectious Diseases* 193 (2006): 112–20.

5. T. R. Dunlap, *DDT: Scientists, Citizens, and Public Policy* (Princeton, NJ: Princeton University Press, 1981), 86–87.

6. C. Lyle, "Achievements and Possibilities in Pest Eradication," *Journal of Economic Entomology* 40 (1947): 1–8.

7. Dunlap, *DDT*, 76–77.

8. E. Conis, "DDT Disbelievers: Health and the New Economic Poisons in Georgia after World War II," *Southern Spaces*, October 28, 2016; https://

southernspaces.org/2016/ddt-disbelievers-health-and-new-economic
-poisons-georgia-after-world-war-ii/.

9. R. A. M. Case, "Toxic Effects of 2,2-bis (p-chlorophenyl)
    1,1,1-trichlorethane (DDT) in Man," *British Medical Journal* 2 (1945):
    842–45.
10. E. Conis, "Beyond *Silent Spring*: An Alternate History of DDT," *Distilla-
    tions* (Science History Institute), February 14, 2017.
11. P. B. Dunbar, "The Food and Drug Administration Looks at Insecti-
    cides," *Food, Drug, Cosmetic Law Quarterly* 4 (1949): 233–39.
12. Dunlap, *DDT*, 135.
13. Dunbar, "Food and Drug Administration."
14. K. Z. Graham, "Federal Regulation of Pesticide Residues: A Brief His-
    tory and Analysis," *Journal of Food Law and Policy* 15 (2019): 98–130.
15. C. Cottam and E. Higgins, "DDT and Its Effect on Fish and Wildlife,"
    Journal of Economic Entomology 39 (1946): 44–52.
16. "Women Advance Wildlife Studies," *New York Times*, May 6, 1945.
17. "DDT Rids Boats of Barnacles," *New York Times*, July 30, 1946.
18. Dunlap, *DDT*, 79–81.
19. G. J. Wallace, "Another Year of Robin Losses on a University Campus,"
    Audubon Magazine 62 (1960): 66–69.
20. R. J. Barker, "Notes on Some Ecological Effects of DDT Sprayed on
    Elms," *Journal of Wildlife Management* 22 (1958): 269–74.
21. Barker, "Ecological Effects of DDT."
22. L. Lear, *Rachel Carson: Witness for Nature*, reprint edition (Boston, MA:
    Houghton Mifflin Harcourt 2009).
23. Lear, *Rachel Carson*.
24. Lear, *Rachel Carson*.
25. Lear, *Rachel Carson*.
26. Lear, *Rachel Carson*.
27. Lear, *Rachel Carson*.
28. Lear, *Rachel Carson*.
29. "Silent Spring Is Now Noisy Summer," *New York Times*, July 22, 1962.
30. Z. Wang, "Responding to *Silent Spring*: Scientists, Popular Science
    Communication, and Environmental Pin the Kennedy Years," *Science
    Communication* 19 (1997): 141–63.
31. R. H. Lutts, "Chemical Fallout: Rachel Carson's *Silent Spring*, Radioac-
    tive Fallout, and the Environmental Movement," *Environmental Review*
    9 (1985): 21–225.
32. F. O. Kelsey, "Thalidomide Update: Regulatory Aspects," *Teratology* 38
    (1988): 221–26.
33. "Thalidomide," in Molecule of the Week archive, American Chemical
    Society (2014); https://www.acs.org/molecule-of-the-week/archive/t
    /thalidomide.html.

34. R. Carson, "A Fable for Tomorrow," *Silent Spring* (Boston: Houghton Mifflin, 1962).

35. K. Walker and L. Walsh, "No One Yet Knows What the Ultimate Consequences May Be": How Rachel Carson Transformed Scientific Uncertainty into a Site for Public Participation in *Silent Spring*," *Journal of Business and Technical Communication* 26 (2012): 3–34.

36. Carson, *Silent Spring.*

37. Carson, *Silent Spring.*

38. Carson, *Silent Spring.*

39. P. Redford, "Counting Our Eagles," *Atlantic*, July 1965; https://www.theatlantic.com/magazine/archive/1965/07/counting-our-eagles/660784/.

40. Carson, *Silent Spring.*

41. Carson, *Silent Spring.*

42. R. Carson, Letter of June 13, 1962, in R. Carson, D. Freeman, and M. E. Freeman, *Always, Rachel: The Letters of Rachel Carson and Dorothy Freeman, 1952–1964* (Boston: Beacon Press, 1995).

43. L. Milne and M. Milne, "There's Poison All Around Us Now" (review of *Silent Spring*), *New York Times Book Review*, September 23, 1962; https://archive.nytimes.com/www.nytimes.com/books/97/10/05/reviews/carson-spring.html?_r=1.

44. W. J. Darby, "Silence, Miss Carson" (review of *Silent Spring*), *Chemical and Engineering News*, October 1, 1962, 60–63.

45. "Pesticides: The Price for Progress" (review of *Silent Spring*), *Time*, September 28, 1962.

46. "Pesticides: The Price for Progress."

47. Carson, *Silent Spring.*

48. Wang, "Responding to *Silent Spring.*"

49. Wang, "Responding to *Silent Spring.*"

50. Wang, "Responding to *Silent Spring.*"

51. Lear, *Rachel Carson.*

52. "Pesticides Report, 15 May 1963," Papers of John F. Kennedy, Presidential Papers, President's Office Files, Departments and Agencies, President's Science Advisory Committee (PSAC); https://www.jfklibrary.org/asset-viewer/archives/JFKPOF/087/JFKPOF-087-003#folder_info.

53. Wang, "Responding to *Silent Spring.*"

54. Lear, *Rachel Carson.*

55. Lear, *Rachel Carson.*

56. "Congress Weighs Stronger Controls on Pesticides," in *CQ Almanac 1964*, 20th ed., (Washington, DC: *Congressional Quarterly*, 1965), 139–41.

57. E. Diamond, "The Myth of the "Pesticide Menace," *Saturday Evening Post*, September 28, 1963.

58. R. H. White-Stevens, "Communications Create Understanding," *Agricultural Chemicals* 17 (1962); reproduced in *DDT, Silent Spring, and the*

Rise of Environmentalism: Classic Texts, ed. Thomas R. Dunlap (Seattle, WA: University of Washington Press, 2008), 109–14.

59. T. R. Dunlap, "Interview with Joseph J. Hickey," in *DDT, Silent Spring, and the Rise of Environmentalism: Classic Texts*, ed. Thomas R. Dunlap (Seattle: University of Washington Press, 2008), 80–84.

60. "U.S.-Owned Lands Curb Pesticides: Udall Issues New Rules on Long-Lived Chemicals," *New York Times*, September 6, 1964.

61. N. Cheremsinoff and P. E. Rosenfeld, eds., *Handbook of Pollution Prevention and Cleaner Production: Best Practices in the Agrochemical Industry* (Norwich, NY: William Andrew, 2011), 247–59.

62. Dunlap, "Interview with Joseph J. Hickey."

63. D. A. *Bird Study* 10 (1963): 56–90.

64. Ratcliffe, "Status of the Peregrine."

65. Dunlap, "Interview with Joseph Hickey."

66. Dunlap, "Interview with Joseph Hickey."

67. D. A. Ratcliffe, "Changes Attributable to Pesticides in Egg Breakage Frequency and Eggshell Thickness in Some British Birds," *Journal of Applied Ecology* 7 (1970): 67–115.

68. L. F. Stickel and L. I. Rhodes, "The Thin Eggshell Problem," USGS report 5210176 (1970). https://pubs.er.usgs.gov/publication/5210176.

69. D. Peakall, "Pesticide-Induced Enzyme Breakdown of Steroids in Birds," *Nature* 216 (1967): 505–6.

70. J. C. Dacre and D. Scott, "Possible DDT Mortality in Young Rainbow Trout," *New Zealand Journal of Marine and Freshwater Research* 5 (1971), 58–65.

71. D. Pimentel, J. Krummel, D. Gallahan, J. Hough, A. Merrill, I. Schreiner, P. Vittum, et al., "Benefits and Costs of Pesticide Use in U. S. Food Production," *BioScience* 28 (1978); 772–84.

72. Pimentel et al., "Pesticide Use in U. S. Food Production."

73. C. Wurster, *DDT Wars: Rescuing Our National Bird, Preventing Cancer, and Creating the Environmental Defense Fund*, illustrated ed. (Oxford, UK: Oxford University Press, 2015).

74. Wurster, *DDT Wars*.

75. Wurster, *DDT Wars*.

76. Wurster, *DDT Wars*.

77. Wurster, *DDT Wars*.

78. See https://www.edf.org/get-involved.

79. Wurster, *DDT Wars*.

80. T. R. Dunlap, "DDT on Trial: The Wisconsin Hearing, 1968–1969," *Wisconsin Magazine of History* 62 (1978): 2–24.

81. W. G. Moore, "The Wisconsin Ban on DDT: Old Law, New Content," *Gargoyle* 16, no. 2 (1985): 3–7 (University of Wisconsin Law School

Forum) https://gargoyle.law.wisc.edu/wp-content/uploads/sites/1169 /2014/02/volume16issue2.pdf.

82. Dunlap, "DDT on Trial."

83. Wurster, *DDT Wars*.

84. Dunlap, "DDT on Trial."

85. W. Hayes, "Sweden Bans DDT," *Archives of Environmental Health* 18 (1969): 872.

86. L. Lundqvist, "Sweden's Environmental Policy," *Ambio* 1 (1972): 90–101.

87. "Reining in the Bug Killer." *Science News* 94 (1968): 337–38.

88. W. D. Ruckelshaus, "Interview by Dr. Michael Gorn," U.S. Environmental Protection Agency Oral History (EPA 202-K-92–0003), January 1993.

89. R. Gillette, "DDT: Its Days Are Numbered, Except Perhaps in Pepper Fields," *Science* 176 (1972): 1313–14.

90. M. G. Harvey, "The Marketing of Potentially Toxic Pesticides Worldwide: The Issues and a Proposed Control System," *Journal of Public Policy & Marketing* 7 (1988): 203–18.

91. A. T. Mhlanga and T. J. Madviza, "Pesticide Residues in Lake McIlwaine, Zimbabwe," *Ambio* 19 (1990): 368–72.

92. G. Matthew, R. S. Yadav, A. S. da Silva, M. Zaim, et al., "Status of Legislation and Regulatory Control of Public Health Pesticides in Countries Endemic with or at Risk of Major Vector-Borne Diseases," *Environmental Health Perspectives* 119 (2011): 1517- 22.

93. P. L. Lallas, "The Stockholm Convention on Persistent Organic Pollutants," *American Journal of International Law* 95 (2001): 692–708.

94. UN Stockholm Convention, Status of ratification (2022); http://chm .pops.int/Countries/StatusofRatifications/PartiesandSignatoires /tabid/4500/Default.aspx.

95. H. van den Berg, G. Manuweera, and F. Konradsen, "Global Trends in the Production and Use of DDT for Control of Malaria and Other Vector-Borne Diseases," *Malaria Journal* 16 (2017): 401.

96. J. Raloff, "The Case for DDT," *Science News* 158 (2000): 12–14.

97. Raloff, "The Case for DDT."

98. S. Basu, "India opposes 2020 deadline for DDT ban"; https://www.downtoearth.org.in/news/ india-opposes-2020-deadline-for-ddt-ban-40967.

99. For the United States in 1963, see Cheremsinoff and Rosenfeld, *Handbook of Pollution Prevention*; for the world in 2014, see Van den Berg, Manuweera, and Flemming, "Global Trends."

100. R. Ritter, M. Scheringer, M. MacLeod, and K. Hungerbuehler, "Assessment of Nonoccupational Exposure to DDT in the Tropics and the

North: Relevance of Uptake via Inhalation from Indoor Residual Spraying," *Environmental Health Perspectives* 119 (2011): 707–12.

101. B. Bienkowski, "DDT Still Killing Birds in Michigan," *Scientific American* Environmental Health News, July 28, 2014.
102. H. N. Geisz, R. M. Dickhut, M. A. Cochran, W. R. Fraser, and H. W. Ducklow, "Melting Glaciers: A Probable Source of DDT to the Antarctic Marine Ecosystem," *Environmental Science and Technology* 42 (2008): 3958–62.
103. M. L. Hladek, A. R. Main, and D. Goulson "Environmental Risks and Challenges Associated with Neonicotinoid," *Environmental Science and Technology* 52 (2018): 3329–35.
104. "EPA Cancels Registrations for 12 Neonicotinoid Pesticides," *Scientist*, May 31, 2019; https://www.the-scientist.com/news-opinion/epa-cancels-registrations-for-12-neonicotinoid-pesticides-65956.
105. R. van Klink, D. E. Bowler, K. B. Gongalsky, A. B. Swengel, A. Gentile, and J. M. Chase, "Meta-Analysis Reveals Declines in Terrestrial but Increases in Freshwater Insect Abundances," *Science* 368 (2020): 417–20.
106. M. J. Davoren and R. H. Schiestl, "Glyphosate-Based Herbicides and Cancer Risk: A Post-IARC Decision Review of Potential Mechanisms, Policy and Avenues of Research," *Carcinogenesis* 39 (2018): 1207–15.
107. C. Davenport, "EPA to Block Pesticide Tied to Neurological Harm in Children," *New York Times*, April 18, 2021.
108. B. A. Cohn, P. M. Cirillo, and M. B. Terry, "DDT and Breast Cancer: Prospective Study of Induction Time and Susceptibility Windows," *Journal of the National Cancer Institute* 111 (2019): 803–10.
109. J. L. Jorgenson, "Aldrin and Dieldrin: A Review of Research on Their Production, Environmental Deposition and Fate, Bioaccumulation, Toxicology, and Epidemiology in the United States," *Environmental Health Perspectives* 109 (2001): 113–39.

CHAPTER SIX

1. M. McCormick, L. Thomason, and C. Trepte, "Atmospheric Effects of the Mt. Pinatubo Eruption," *Nature* 373 (1995): 399–404.
2. J. Hansen, A. Lacis, R. Ruedy, and M. Sato, "Potential Climate Impact of Mount Pinatubo Eruption," *Geophysical Research Letters* 19 (1992): 215–18.
3. R. Campion, "The Eruption of Mount Tambora," *Thomas Jefferson's Monticello*, June 1, 2021; https://www.monticello.org/site/research-and-collections/eruption-mount-tambora.

4. United Nations Conference on Environment and Development, Rio de Janeiro, Brazil, 3–14 June 1992; https://www.un.org/en/conferences /environment/rio1992.

5. R. Jackson, "Eunice Foote, John Tyndall and a Question of Priority," *Royal Society Journal of the History of Science: Notes and Records* 74 (2020): 105–18.

6. M. Thatcher, "Speech at 2nd World Climate Conference," November 6, 1990, available at Margaret Thatcher Foundation; https://www.marga retthatcher.org/document/108237).

7. B. P. Perry, "How Thatcher Broke the Miners' Strike But at What Cost?" Online at https://www.history.co.uk/article/how-thatcher-broke-the -miners-strike-but-at-what-cost.

8. M. Paterson and M. Grubb, "The International Politics of Climate Change," *International Affairs* 68 (1992): 293–310.

9. Paterson and Grubb, "International Politics of Climate Change."

10. R. Rapier, "The 10 Countries That Dominate World Fossil Fuel Production," *Forbes*, July 14, 2019; https://www.forbes.com/sites/rrapier /2019/07/14/ ten-countries-that-dominate-fossil-fuel-production/?sh=65feed0e5b13.

11. Paterson and Grubb, "International Politics of Climate Change."

12. J. Pal and E. Eltahir, "Future Temperature in Southwest Asia Projected to Exceed a Threshold for Human Adaptability," *Nature Climate Change* 6 (2016): 197–200.

13. N. Swart and A. Weaver, "The Alberta Oil Sands and Climate," *Nature Climate Change* 2 (2012): 134–36.

14. S. C. Sherwood and M. Huber, "An Adaptability Limit to Climate Change Due to Heat Stress," *Proceedings of the National Academy of Sciences* 107 (2010): 9552–55.

15. C. F. Schleussner, J. Rogelj, M. Schaeffer, T. Lissner, et al., "Science and Policy Characteristics of the Paris Agreement Temperature Goal," *Nature Climate Change* 6 (2016): 827–35.

16. M. R. Raupach, G. Marland, P. Ciais, C. Le Quéré, J. G. Canadell, and G. Klepper, and C. B. Field, "Global and Regional Drivers of Accelerating CO_2 Emissions," *Proceedings of the National Academy of Sciences* 104 (2007): 10288–93.

17. M. W. Browne, "Grappling with the Cost of Saving Earth's Ozone," *New York Times*, July 17, 1990.

18. K. Bond, E. Vaughan, and H. Benham, "Decline and Fall: The Size and Vulnerability of the Fossil Fuel System," Report by Carbon Tracker, June 2020; https://carbontracker.org/reports/decline-and-fall/.

19. N. Oreskes and E. M. Conway, *Merchants of Doubt: How a Handful of Scientists Obscured the Truth on Issues from Tobacco Smoke to Global Warming* (Bloomsbury Publishing, 2011).

20. A. Robock, "Benefits and Risks of Stratospheric Solar Radiation Management for Climate Intervention (Geoengineering)," *The Bridge* 50 (2020): 59–67.

21. S. Barrett, "Solar Geoengineering's Brave New World: Thoughts on the Governance of Unprecedented Technology," *Review of Environmental Economics and Policy* 8 (2014): 246–69.

22. C. Okereke, "My Continent Is Not Your Giant Climate Laboratory," *New York Times*, April 18, 2023.

23. S. Ungar, "Bringing the Issue Back In: Comparing the Marketability of the Ozone Hole and Global Warming," *Social Problems* 45 (1998): 510–27.

24. G. A. Meehl, C. Tebaldi, and D. Adams-Smith, "US Daily Temperature Records Past, Present, and Future," *Proceedings of the National Academy of Sciences* 113 (2016): 13977–82.

25. J. Aldy, "Green New Deal Can Learn from Obama's $90bn Clean Energy Program of 2009," *energypost.eu*, February 22, 2019; https://energypost.eu/green-new-deal-can-learn-from-obamas-90bn-clean-energy-plan-of-2009/.

26. For inflation reduction, see M. Barbanell, "A Brief Summary of the Climate and Energy Provisions of the Inflation Reduction Act of 2022," World Resources Institute, October 28, 2022. https://www.wri.org/update/brief-summary-climate-and-energy-provisions-inflation-reduction-act-2022. On infrastructure, see The White House, Fact Sheet at https://www.whitehouse.gov/briefing-room/statements-releases/2022/11/15/fact-sheet-one-year-into-implementation-of-bipartisan-infrastructure-law-biden-%E2%81%A0harris-administration-celebrates-major-progress-in-building-a-better-america/.

27. F. Harvey, "COP 21: Paris Climate Change Conference 2015: Paris Climate Change Agreement: The World's Greatest Diplomatic Success," *Guardian*, December 14, 2015.

28. United Nations, Text of the Paris Agreement; https://unfccc.int/process/conferences/pastconferences/paris-climate-change-conference-november-2015/paris-agreement.

29. P. Eisenstein, "GM to Go All-Electric by 2035, Phase-Out Gas and Diesel Engines," *NBC News*, January 28, 2021; https://www.nbcnews.com/business/autos/gm-go-all-electric-2035-phase-out-gas-diesel-engines-n1256055.

30. C. Crownhart, "What's Next for Batteries?" *MIT Technology Review*, January 4, 2023.

31. M. Bergen and H. Mountford, "Five Ways Momentum for Climate Action Has Grown Since The Paris Agreement Was Signed," *World Resources Institute*; https://www.wri.org/insights/6-signs-progress-adoption-paris-agreement, September 21, 2020.

32. European Commission, "Communication from the Commission: The European Green Deal," Brussels, 11.12.2019 COM(2019) 640 final; https://eur-lex.europa.eu/resource.html?uri=cellar:b828d165-1c22 –11ea-8c1f-01aa75ed71a1.0002.02/DOC_1&format=PDF.
33. Barbanell, "Inflation Reduction Act of 2022."
34. Encyclical letter (The Holy See), "'LaudatoSi' of the Holy Father Francis: On Care for Our Common Home," 2015; https://www.vatican.va/content/francesco/en/encyclicals/documents/papa-francesco_20150524_enciclica-laudato-si.html.
35. S. Haynes, "Students from 1,600 Cities Just Walked Out of School to Protest Climate Change: It Could Be Greta Thunberg's Biggest Strike Yet," *Time*, May 24, 2019; https://time.com/5595365/global-climate-strikes-greta-thunberg/.
36. J. Brady, "In a Landmark Case, A Dutch Court Orders Shell to Cut Its Carbon Emissions Faster," *NPR*, May 26, 2021; https://www.npr.org/2021/05/26/1000475878/in-landmark-case-dutch-court-orders-shell-to-cut-its-carbon-emissions-faster.
37. "State Suits Against Oil Companies," *State Energy and Environmental Impact Center, NYU School of Law*; https://stateimpactcenter.org/issues/climate-action/suits-against-oil-companies.
38. D. Gelles, D. "In Montana, It's Youth versus the State in a Landmark Climate Case," *New York Times*, March 21, 2023.
39. "Engine No. 1 Extends Gains with a Third Seat on Exxon Board," *Reuters*, June 2, 2021; https://www.reuters.com/business/energy/engine-no-1-win-third-seat-exxon-board-based-preliminary-results-2021-06-02/.
40. W. Bauck, "Where the Sunrise Movement Goes from Here," *Slate*, November 28, 2022; https://slate.com/news-and-politics/2022/11/sunrise-movement-biden-ira-2024.html#:~:text=Sunrise%20officially%20responded%20by%20saying,to%20approach%20the%20organization%27s%20next.
41. S. Hsiang, R. Kopp, A. Jina, J. Rising, M. Delgado, et al., "Estimating Economic Damage from Climate Change in the United States," *Science* 356 (2017): 1362–69.
42. J. R. Marlon, X. Wang, P. Bergquist, P. D. Howe, A. Leiserowitz, E. Maibach, M. Mildenberger, and S. Rosenthal, "Change in U.S. State-Level Public Opinion about Climate Change, 2008–2020," *Environmental Research Letters* 17 (2022). https://doi.org/10.1088/1748-9326/aca702.
43. E. Nilsen, "Even before Ukraine Crisis, Majority of Americans Wanted Country to Prioritize Renewable Energy Development, Poll Shows," *CNN*, March 1, 2022; https://www.cnn.com/2022/03/01/politics/renewable-energy-poll-ukraine-climate/index.html.

ILLUSTRATION CREDITS

p. 17   From R. Benedick, *Ozone Diplomacy* (Cambridge, MA: Harvard University Press, 1991), 27. Copyright 1991, 1998 by the President and Fellows of Harvard College. Used by permission. All rights reserved.

p. 28   Courtesy of NASA Ozone Watch, NASA/Goddard Space Flight Center Scientific Visualization Studio, Greg Shirah, animator, and Paul Newman, scientist. Available at https://svs.gsfc.nasa.gov/718.

p. 34   Photograph by David J. Hofmann.

p. 59   Courtesy of National Archives, photo no. 26-G-3422. Series: American Unofficial Collection of World War I Photographs, 1860—1952. Available at https://catalog.archives.gov/id/533758.

p. 70   Adapted from Scialabba, N., et al. Food wastage footprint: impacts on natural resources, summary report. Food and Agriculture Organization of the United Nations, ISBN 978-92-5-107752-8, 2013; updated at https://www.fao.org/documents/card/en/c/7338e109 -45e8-42da-92f3-ceb8d92002b0/.

p. 93   Courtesy of California Institute of Technology Archives and Special Collections.

p. 100   Courtesy of University of Southern California, on behalf of USC Libraries Special Collections, USC Digital Library, Los Angeles Examiner Photographs Collection.

p. 145   Adapted from "The Nature and Extent of Lead Poisoning in Children in the United States: A Report to Congress." Agency for Toxic Substances and Disease Registry, July 1988. Online at https://stacks .cdc.gov/view/cdc/13238.

p. 161   Photograph by Alfred G. Etter, published with permission from the Alfred G. Etter Estate.

p. 175   Courtesy of the State Library and Archives of Florida/ Joseph Janney Steinmetz.

p. 182   Courtesy of the Associated Press, AP Photo/Charles Gorry.

p. 217   Courtesy of the National Mining Association.

p. 219   Data from Hannah Ritchie, Max Roser and Pablo Rosado "CO_2 and Greenhouse Gas Emissions" (2020), published online at OurWorld InData.org: https://ourworldindata.org/emissions-drivers. Figure by the author.

p. 222   Adapted from D. M. Etheridge, L. P. Steele, R. L. Langenfelds, R. J. Francey, J. M. Barnola, and V. I. Morgan, "Historical CO_2 Records from the Law Dome DE08, DE08-2, and DSS Ice Cores (1006 AD–1978 AD)," (Carbon Dioxide Information Analysis Center [CDIAC], 1998), Oak Ridge National Laboratory (ORNL), Oak Ridge, TN (United States), ESS-DIVE repository. Dataset. doi:10.3334/CDIAC/ATG.011; and Dr. Pieter Tans, NOAA/GML (gml.noaa.gov/ccgg/trends/) and Dr. Ralph Keeling, Scripps Institution of Oceanography (scrippsco2.ucsd.edu/). (*inset*) "Global Averaged Land and Ocean Temperature Anomalies, Relative to the Average over the Full Period," NOAA National Centers for Environmental Information, Climate at a Glance: Global Time Series, published June 2023, data available online at https://www.ncei.noaa.gov/access/monitoring/climate-at-a-glance/global/time-series. Figure by the author.

p. 225   Adapted from M. R. Raupach, G. Marland, P. Ciais, C. Le Quéré, J. G. Canadell, and G. Klepper, and C. B. Field, "Global and Regional Drivers of Accelerating CO_2 Emissions," *Proceedings of the National Academy of Sciences* 104 (2007): 10288–93. Used with permission. Copyright (2007) National Academy of Sciences, USA.

p. 241   1883 Photo: USGS Photo/Israel Russell. 2015 Photo: NPS Photo/Keenan Takahashi. Online at https://www.nps.gov/articles/glaciersandclimatechange.htm.

p. 247   Courtesy of Lazard LCOE+ Analysis. From https://www.lazard.com/research-insights/2023-levelized-cost-of-energyplus/.

p. 258   Images courtesy of Greenpeace and photographers Phillip Reynaers (top) and Pierre Gleizes (bottom).

INDEX

air pollution and smog (*cont.*)
and hydrocarbons, 88–92, 108–12;
and industrialization, 81, 115, 211;
legislation, 20, 83–84, 93, 98, 103–4,
109, 144; local and state, 79–90,
93–96, 101–5, 108–14, 119, 137, 143;
major components of, 81, 88–92,
106–19, 143–44, 209; measurement
of, 83; and metals, 83; nationwide,
94, 103–6, 109, 119, 137, 144; and
nitrogen oxide, 81, 91, 106, 108,
111–15, 143–44; as nuisance, 82,
84; and ozone depletion, 8, 254;
photochemical generation of, 92;
and policy, 85–86, 120; in poor
neighborhoods and countries, 119–
20; protest against (Los Angeles,
1961), 100; and public health, 79, 83,
102, 108, 118; and public opinion,
101; and redlining in nonwhite
neighborhoods, 119; and soot, 228;
study (Los Angeles), 86; and sulfur
dioxide, 81, 89, 108, 114–19, 209;
and technology for cleanup, ix, 120;
term, usage, 81; US as world leader
in cleaning up, 119; and vegetation
damage, 89–90, 115; widespread, 139;
and zinc, 82–83. *See also* acid rain
Air Resources Board (California), 105
aldrin insecticide, 198–201, 204, 278n109
Alliance for Responsible Atmospheric
Policy, 69, 72
Alliance for Responsible CFC Policy,
34–35
Alliance of Small Island States, and
climate change, 218
American Chemical Society (ACS), 8–9
American Clean Energy and Security
Act of 2009 (ACES), 72
American Cyanamid Company, 179, 183
American Medical Association, 179
American Motors, 106
American Psychological Association
(APA), x
American Recovery and Reinvestment
Act of 2009, 242
American Steel and Wire, 82–83

Anderson, Jack, 107
Anderson, Jim, 33
Andes, climate change in, 240
Antarctic: chlorine chemistry
measurements, 32–33; and climate
change, viii; ozone depletion in, 2–3,
27, 29–34, 42, 45–46, 49–51, 221, 238
Antarctica: climate change in, 235,
239; DDT in, 201–2; dryness and
humidity in, 235; and lead in ice and
snow, 133; McMurdo Station, 31–32;
ozone depletion in, 8, 21, 27–32;
pesticides in, 201–2; shifting ice
floes in, 1, 239; stratosphere, 29–30.
See also South Pole
APA. *See* American Psychological
Association (APA)
Arctic: and climate change, 248; ozone
depletion in, 2, 29, 31, 45–46, 49.
See also North Pole
Asia: air pollution and smog in, 112;
climate change and global warming
in, 241; nitrogen oxide emissions in,
112; pesticides in, 199
Atlantic Monthly, 166
atmospheric chemistry: and climate
change, 211, 213–14; and greenhouse
gases, 55–57, 62; and ozone depletion,
2, 7, 22–23, 43, 45–46. *See also*
stratospheric chemistry
atomic bomb, 152, 181–82
Atomic Energy Commission (AEC),
132–33
Attenborough, David, 167
Audubon Magazine, 160–61, 161, 274n19
Audubon Society, 167–68, 175–77, 191
Australia: carbon dioxide emissions in,
70, 218, 219; and climate change,
218; greenhouse gas emissions in,
70; lead in, 126, 129; and ozone
depletion, 40, 42; pesticides in, 196
auto industry: and air pollution/smog,
ix, 93–94, 102–14, 119, 145, 230; and
climate change, 230, 244–45, 249;
and fuel economy, 109, 111–12, 244;
and greenhouse gases, 62–63; and
lead in air, 137–38, 143–45; and

fungicides, 158, 188, 202. *See also* herbicides; insecticides; pesticides
FWS. *See* US Fish and Wildlife Service (FWS), and DDT effects

Garcia, Rolando, 26, 29
Garwin, Richard, 178–79, 182–83
gas chromatography, electron capture, 8. *See also* vapor phase chromatography
gas industry. *See* oil and gas industries
Gates Foundation, 200–201
General Motors (GM), 63, 83, 106, 110, 123, 244
geochemistry, and lead, 132, 134
geoengineering, and climate change, 236–38
Germany: carbon dioxide emissions in, 70, 219; CFCs and ozone depletion in, 20, 44; greenhouse gas emissions in, 70; lead in, 125; pesticides in, 186, 196; thalidomide drug use in, 171
Gibson, J. Lockhart, 126–27, 129, 148–50, 271n16
glaciers: ice in, 201; retreat of, 240–41, 241, 256; structure and motion of, 212
global north, and climate change, 214, 218, 227–28
global south, and climate change, 214, 218, 221, 227–28
global warming, 207–8, 220–21, 234–35, 240–41, 247–48; and carbon dioxide, 212, 222, 235, 254; dangers and grave threat of, 208, 214–15; and fossil fuels, 208, 220; and greenhouse gases, 54, 57, 61–62, 66, 69–72. *See also* climate change
glyphosate herbicide, 202–3
GM. *See* General Motors (GM)
Goddard Institute for Space Studies, 209
Goldberg, Herbert, 171
Gore, Al, 19, 57
Gorsuch, Anne, 23–24
government: and environment, 196; and industry, 7, 22, 61, 104, 252; and public will, 104. *See also* politics

Great Britain, pesticides in, 180. *See also* Britain; England, lead in; United Kingdom (UK)
Great Lakes, and ice age, 234
greenhouse gases, 53–78; alternatives to, 254; and atmospheric chemistry, 55–57, 62; and auto industry, 62–63; and carbon dioxide, 55, 58, 62, 74, 210–11, 228, 234; and chemical industry, 60–62, 67–72, 78; and climate change, 210–11, 228, 234, 246–49, 254; emissions, 61, 64–65, 73–74, 228, 234, 246–49, 254; emissions, global (2011), 70; and food loss/waste, 68, 70; and global warming, 54, 57, 61–62, 66, 69–72; heat by infrared light absorbed by, 210; and hydrocarbons, 74; and ozone depletion, 44, 53; reductions in, 67, 247; as rich countries responsibility, 74. *See also* carbon dioxide; fossil fuels; hydrofluorocarbons (HFCs); methane gas
Greenland: and climate change, 239; and lead in ice and snow, 125, 133–36, 148, 270n9; shifting ice floes in, 239
Greenpeace, 47, 222, 252, 257–58; activists in Paris, 258
Griffin, Robert P., 113
Griffin-Riegle proposal, 113
Gruening, Ernest Henry, 181
Guatemala, carbon dioxide emissions in, 219

Haagen-Smit, Arie Jan (Haggy), 87–93, 93, 105
hairsprays: advertisement (1950s), 12; CFCs in and ozone depletion, 11, 13, 18. *See also* deodorants
halons, in fire extinguishers, and ozone depletion, 20, 45
Hamilton, Alice, 127–28, 148–50, 271nn17–19
Hansen, James, 209
Hartley, Walter, 3–4
Hawaii, chlorpyrifos banned in, 203

Hayes, Denis, 98, 101
HCFCs. *See* hydrochlorofluorocarbons (HCFCs)
health. *See* public health
heavy metals. *See under* metals
heptachlor insecticide, 156, 199
herbicides, 97, 188, 202–3. *See also* fungicides; insecticides; pesticides
HEW. *See* US Dept. of Health, Education, and Welfare (HEW)
HFCs. *See* hydrofluorocarbons (HFCs)
HFOs. *See* hydrofluoro-olefins (HFOs), and refrigeration
Hickey, Joseph, 184–86, 194
Higgins, Elmer, 159, 165–67, 274n15
Hodel, Don, 38–39
Human Environment Conference, 211
human health. *See* public health
hunger, global, 69, 238–39
hurricanes, 239, 241
hydrocarbons: in gasoline, 90; and greenhouse gases, 74; and refrigeration, 74; smog-producing, 88–92, 108–12. *See also* methane gas
hydrochlorofluorocarbons (HCFCs): and greenhouse gases, 61; and ozone depletion, 43–49
hydrofluorocarbons (HFCs): alternatives to, 63, 70–71, 77; and greenhouse gases, 57–58, 61, 63, 67, 70–77; and ozone depletion, 43–47; phaseout, reduction, and bans of, 58, 61, 63, 67, 72–76; regulation of, 76
hydrofluoro-olefins (HFOs), and refrigeration, 62, 70, 74

ice ages, 233–34
Illinois: air pollution and smog in, 118; pesticides in, 160–61, 173–75, 178
Illinois Agricultural Experiment Station, 178
India: air pollution and smog in, 120; carbon dioxide emissions in, 70, 219; CFCs phaseout and bans in, 68; and climate change, 227, 248; DDT use in, 200; fossil fuel emissions

in, 227; greenhouse gases in, 68, 70, 70–71, 73–74, 76, 248; and ozone depletion, 41
Indiana, air pollution and smog in, 118
Indonesia: carbon dioxide emissions in, 70, 218; greenhouse gas emissions in, 70; pesticides in, 183
industrialization: and air pollution/smog, 81, 115, 211; and carbon dioxide, 211
industrial toxicology, 127–28
industry: and environment, 17–18, 75, 78, 205; and government, 7, 22, 61, 104, 252; heavy, 119; and information asymmetry, between regulators and manufacturers, 109, 115; and innovation, 109; and law, 157; and public health, 137; and science, 16, 28–29, 188, 195; and technological change, ix. *See also specific industries*
inequality, and discrimination, 140–42, 149. *See also* environmental justice; racial justice; social justice
Inflation Reduction Act, 242, 252–53
information asymmetry, between regulators and manufacturers, 109, 115
infrared light, 55, 210, 212, 234. *See also* ultraviolet (UV) sunlight
infrastructure, 48, 119, 149, 249, 252–53
innovation: and industry, 109; and technology, ix, 36, 144, 241
insect-borne diseases, 152–53
insecticides, 153–54, 172–74, 178, 187–90, 199–200; and food production, 156; organochlorine molecules in, 156; regulation of, 156. *See also* DDT; fungicides; herbicides; pesticides; *and specific insecticides*
Intergovernmental Panel on Climate Change (IPCC), 63, 214, 222, 232, 245–46; approval process (2007), 258; Scientific Assessment of Climate Change (2007), 255–57. *See also* United Nations (UN)
International Geophysical Year, 20–21, 133–34

Mount Tambora volcano (Indonesia), 209, 278n3
Muskie, Edmund (Ed), ix, 94–96, 99–107, 110–14, 135–39, 144, 149–50, 241–42, 254
Myanmar dictatorship, and environmental risks to public health, 148

NAACP. *See* National Association for the Advancement of Colored People (NAACP), Legal Defense Fund
NACA. *See* National Agricultural Chemicals Association (NACA)
napalm, and Vietnam War usage, 97
NAS. *See* National Academy of Sciences (NAS)
NASA. *See* National Aeronautics and Space Administration (NASA)
National Academy of Sciences (NAS), 22, 24, 110–11, 169, 198
National Aeronautics and Space Administration (NASA), 19, 26–28, 33, 37, 54, 209
National Agricultural Chemicals Association (NACA), 156
National Association for the Advancement of Colored People (NAACP), Legal Defense Fund, 189
National Center for Atmospheric Research (NCAR), 26
National Coal Association, 23
National Council of Women of the United States (NCW/US), 176
National Lead Company, 129–30
National Marine Fisheries Service (NMFS), 138–39
National Oceanic and Atmospheric Administration (NOAA), 26, 37
National Ozone Expedition (NOzE), 2–3; balloon payload recovery (1986), 34
National Parks Association, 176
National Science Foundation (NSF), 6, 133–34
National Wildlife Federation (NWF), 167

Natural Resources Defense Council (NRDC), 19, 24–25, 109
nature: and environment, 189; and policy, 205; and science, 180
Nature Conservancy Council (NCC), 185
Nature journal, 9, 27
NCAR. *See* National Center for Atmospheric Research (NCAR)
NCC. *See* Nature Conservancy Council (NCC)
NCW/US. *See* National Council of Women of the United States (NCW/US)
Needleman, Herbert, 146–47, 149–50
Nelson, Gaylord, 98, 100, 103, 194
neonicitinoid pesticides, 202
Nepal, carbon dioxide emissions in, 218, 219
Netherlands: carbon dioxide emissions in, 219, 252; CFCs and ozone depletion in, 20, 25, 44, 252; and climate change, 252; DDT use in, 160; Dutch elm disease in, 160; emissions reductions in, 252; pesticides in, 160
Nevada, air pollution and smog in, 104
New England, tides and climate change in, 240–41
New Jersey: air pollution and smog in, 84, 265n13; pesticides in, 159
New York (state): air pollution and smog in, 84, 104; as antipollution state, 104; chlorpyrifos banned in, 203; pesticides in, 155–56, 189–90, 192, 203
New York City (NYC): air pollution and smog in, 94–95, 104; and lead poisoning, 121–23, 126, 141
New Yorker, 166, 169, 178–79
New York Times, 98, 159–60, 166–67, 170, 177, 187, 193–94
New Zealand: carbon dioxide emissions in, 37, 218, 219; CFCs banned in, 37; and climate change, 218; and ozone depletion, 37, 40, 42
Nicaragua, carbon dioxide emissions in, 218, 219

West Germany, CFCs phaseout and
bans in, 20, 44
White-Stevens, Robert H., 183, 275n58
WHO. *See* World Health Organization
(WHO)
Wiesner, Jerome, 179, 182–83
wildfires, 207–8, 220, 239–40, 248
wind: energy, 242, 245–46, 247, 249;
turbines, 229
Wisconsin: air pollution and smog in,
98; pesticides in, 184, 186, 192–98,
276–77nn80–81
"Women Scientists Advance Wildlife
Studies" (Stickel), 159–60, 274n16
women's rights, 142
World Climate Conference (1990),
214–15, 279n6

World Health Organization (WHO),
179, 199–200
World's Fair (Chicago, 1893), 59–60
Wurster, Charles, 189–91, 194
Wyoming, Powder River Basin, clean
low-sulfur coal from, 118

Yannacone, Carol, 190
Yannacone, Victor, Jr., 189–93
Young Lords (NYC street gang), 122–
23, 141, 270nn1–2

zealotry, 48–49
Zero Hunger Challenge (ZHC), 69
zinc industry, 82–83, 129
zinc oxide, as lead alternative, 129
Zinc Works, 82–83